The Brain

10 Things You Should Know

Sophie Scott CBE FBA FMedSci is a British Cognitive Neuroscientist and Director of the Institute of Cognitive Neuroscience, University College London (UCL). She has been researching the human brain since the 1990s, and has published over 130 peer-reviewed papers. Her research investigates the brain bases of human vocal communication, including voices, speech and laughter. She was the 2017 Royal Institution Christmas Lecturer and she was awarded the Royal Society's Michael Faraday prize and lecture in 2021.

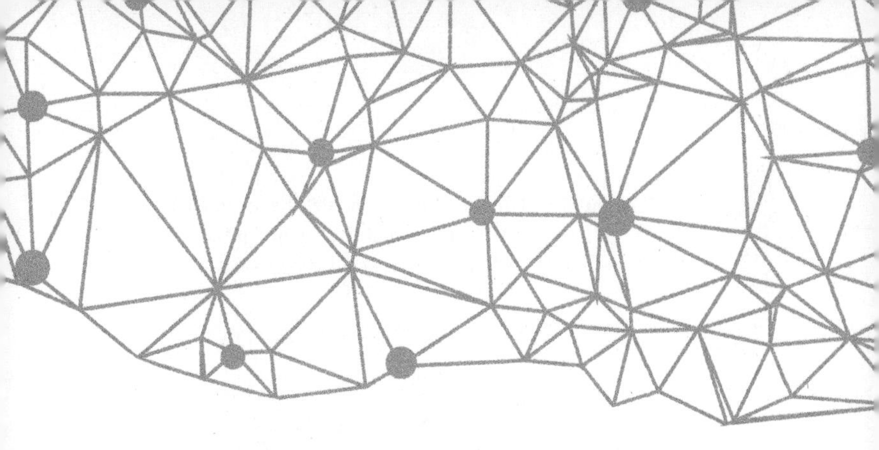

The Brain

10 Things You Should Know

Explore the wonders of our most extraordinary organ

PROFESSOR SOPHIE SCOTT

SEVEN DIALS

First published in Great Britain in 2022 by Seven Dials
an imprint of The Orion Publishing Group Ltd
Carmelite House, 50 Victoria Embankment
London EC4Y 0DZ

An Hachette UK Company

3 5 7 9 10 8 6 4 2

A CIP catalogue record for this book is
available from the British Library.

ISBN (Hardback) 978 1 3996 0292 1
ISBN (eBook) 978 1 3996 0293 8

Printed in Great Britain by Clays Ltd, Elcograf S.p.A.

www.orionbooks.co.uk

Dedication

This book is dedicated to the brilliant brains of Tom Manly, our son Hector, my Mum Christine Scott and my Dad Colin Scott.

Contents

Preface

I love brains. I love your brain. The most interesting part of anyone I meet is, without doubt, their brain. And I am lucky enough to have been in the right place at the right time to get to look at brains for a living. For a lot of human history, brains have not always been considered very interesting – Aristotle thought that the heart was the organ for sensation and experience, and that the brain was for cooling the body. In Aristotle's defence, it does feel like we experience things in our bodies – when I am happy, I feel it in my heart; when I get the first inkling that something is horribly wrong, I feel that sinking feeling in my stomach. Now, of course, we know that everything we experience is our brain's best guess at both what is out there in the world, and what is going on in our bodies. Furthermore we know that the brain maps out many things (like emotions) in the body itself.

Another reason Aristotle thought that the brain was not very important is that when the brain itself is exposed, it has no sensation – if you touch the surface of an exposed brain (please don't), the owner of the brain will not feel that touch. The brain itself has no way of perceiving the world directly – it gathers information from our sensory organs (as we will see in this book), and continually uses this to try and construct our reality.

A further reason why Aristotle and others were unimpressed by the brain was because doing anatomical investigations on the brain was very tricky: in its natural state, the brain is not very solid. It has a texture like a rather unprepossessing grey, watery jelly. Although TV detective programmes like to show the whole brain being taken out and passed around, post-mortem, in reality a variety of different techniques are used to make the brain solid enough to make an anatomical investigation possible.

Not everyone agreed with Aristotle about the brain, and despite the difficulties in working with brains, some impressive discoveries were made. In 500 BC, Alcmaeon of Croton discovered the optic nerve and described phosphenes – visual images that are caused by activity in the optic nerve. Plato wrote that the brain was 'the divinest part of us and master over all the rest' (where 'the rest' are the chest and the liver). However, the growing scientific consensus around how the brain worked focused on the ventricles, not on the brain itself: the brain floats inside the skull, and the fluid in which it floats flows into the gaps between different parts of the brain. These gaps, called ventricles, were considered the seat of the brain's abilities, rather than the manky, gooey tissue surrounding them.

Many more discoveries about the brain were made over the next centuries, including those by Abū Bakr Muhammad Ibn Zakariya al-Rāzī, the Persian scientist, medic and philosopher who in AD 900 determined that nerves either had sensory or motor functions, and was able to relate symptoms to the location of brain damage. But the ventricular theory of the brain still held sway in Western medicine – when Da Vinci drew the brain, he emphasised the ventricles. It was not until 1538 that the influential medic Andreas Vesalius overthrew the ventricular theory, and the concept that it was the brain itself that was critical to mental processes started to dominate Western medicine.

From here, our understanding of the brain developed in pace with technology. The development of the microscope led to the discovery of cells by Robert Hooke in 1665, and the proposal by Theodor Schwann that all living things were made up of cells. The brain, for a long time, seemed to be an exception to this theory – there were certainly cells in there, but what was all this fibrous matter that also seemed to be there in abundance? A scientist named Camillo Golgi invented a staining technique that enabled him to see (using a microscope) that these fine strands were connected to, and formed part of, brain cells (neurones – more on what these are later). Golgi (incorrectly) thought that these projections formed one huge, continuous network. Using the same staining technique, the Spanish neuroscientist Santiago Ramón y Cajal identified that the neurones in the brains of birds were not one large mesh, but made of individual cells – and thus that, like the rest of life on earth, the brain was also made up of individual cells.

During the nineteenth century, there were also big developments in the understanding of how the brain controls behaviour. German and French neurologists mapped out where language processes were found in the left hemisphere, with such accuracy that, although we have developed our understandings of the brain mechanisms involved, we are still working in the same brain areas they described. In the mid-twentieth century Walter Penfield in Montreal performed studies on the exposed brains of patients who were having brain surgery to relieve intractable epilepsy, and showed that gentle electrical stimulation of different brain areas would lead to different kinds of perceptions or actions.

Towards the end of the last century, there was an explosion in the techniques of brain science, meaning that we could make images of brain structure, and examine human brain function, without anyone needing to die or have brain surgery. Techniques like magnetic resonance imaging and EEG/MEG are now standard tools for both clinical investigation of and basic research into the human brain. In my career, I have been honoured to work with patients with brain damage, and to study brains using brain imaging, since the 1990s, and I still consider that I have the best job in the world. I have learned something from every brain I have met. And that is because, as we will see in this book, the brain – your brain – is one of the most miraculous structures in the known universe.

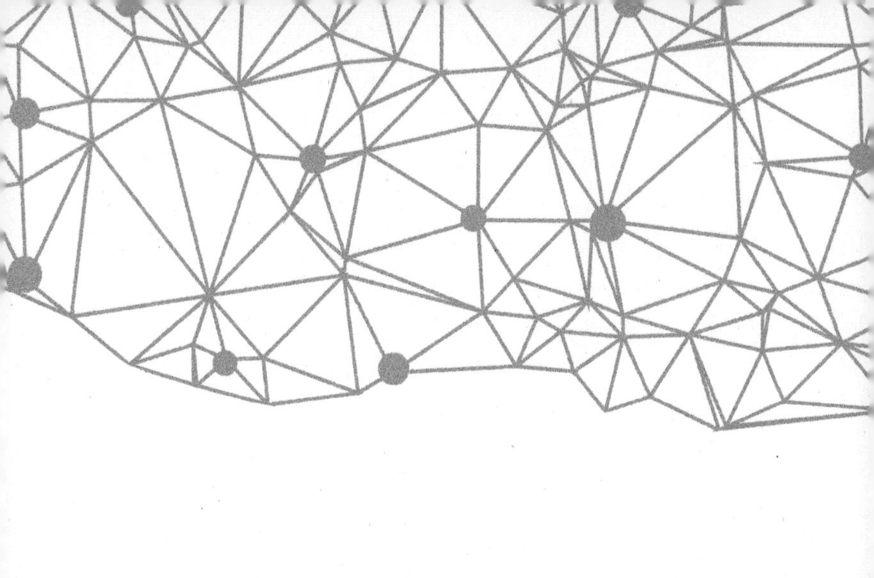

1

Why Are You Still You?

Look at a photo of you as a baby. Almost every part of you in that photo isn't there anymore, because not many things stay with you for your whole life. Hairs fall out, nails grow, cells throughout your body die and are replaced by new ones. This cycle of death and renewal in the body leads to the stunning claim that every ten years or so, you have a whole new body. However, this is not quite true. Two kinds of cells in your body do not die and get replaced. The cells in the lenses of your eyes are with you for life, and so are the nerve cells (neurones) in your central nervous system. Your central nervous system consists of the nerves in your spine and in your brain. It's quite dizzying to think that when each of us was born, we already had almost all the nerve cells in our brains that we would ever have. When you are looking at the photo of you as a baby, you are looking at someone with the same brain cells as you, and a completely different body.

And there are so many of these neurones. The human brain contains some 86 billion neurones, which are highly specialised to power your mental processes.

The neurones in your brain do not form a single mass: they are highly structured into a very complex and distinctive pattern of brain areas, forming huge and extremely complicated networks. When I was an undergraduate first studying the brain, this structure seemed pointlessly, incomprehensively complex. What on earth could justify such impenetrable anatomical detail? It took me a long while to realise that it's because your brain is not just responsible for everything you do, and for storing everything you learn: it is also responsible for everything you experience, everything you remember, everything you look forward to: in short, it's responsible for you being *you*. Indeed, everything you know, every skill you have, every memory you have, every fact you know, is encoded in networks of these neurones. Which is why all those billions of neurones need to stay with you for the long haul – you can change all the cells in the lining of your gut every two to four days, or all your red blood cells every four months, and remain the same person: switch out all your neurones, and you would be *a whole new person*.

Within the brain, neurones form networks, where they can connect with each other and talk to each other. This property underlies the unusual shape of neurones, which look very different to other cells in the body. Obviously, different cells throughout the body are vastly different from one another, in ways that reflect their functional roles: muscle cells look and behave differently from skin cells, for example. Some cells are highly distinct: the hair cells in your ear have moving parts that let them change vibrations into sounds, and red blood cells are flat and flexible, so that oxygen can permeate into them, and so that they can squeeze through tiny capillaries and carry that oxygen around the body. Even among this strange menagerie, however, neurones are quite extraordinary. Some look like trees; some look like cornflowers with long stalks and spikey petals. They bristle with filaments that project from the cell body, some long, some short. The structure of neurones lies at the heart of the neurones' special abilities, allowing them to connect to other neurones, and thus to communicate with each other in large networks.

There are three properties of neurones that mean they can talk to each other in these networks. First, the neurones can form connections with each other. Second, the neurones can be activated. Third, when they are active, they can pass on messages to the other neurones to which they are connected.

We find the connections between neurones when we examine the fine projections that sprout from the cell bodies. The ends of these projections flatten out into little buttons, which abut the flat buttons at the ends of other neurones. The buttons do not quite touch, but have a tiny gap between them. These junctions between neurones are called a synapse; synapses are the ways that neurones form and maintain connections with one another.

When a neurone is activated, or 'fires', it creates an electrical discharge that spreads across the whole neurone and all its projections. In other words, communication within a neurone is caused by electrical signalling. You have probably had a lot of experience with electricity – current electricity in your home that powers your light bulbs and your fridge, and static electricity that makes your hair stand on end when you rub it with a balloon. Both these kinds of electricity rely on the movement of electrons. A third kind of electricity – biological electricity – creates electrical charges across cell membranes by moving ions around (ions are atoms or molecules that carry an electrical charge). In our neurones, an electrical charge is built up by changing the concentrations of sodium and potassium ions on either side of the cell membranes that coat the neurones.

The electrical message moves extremely quickly from one end of the neurone to the other – across the short projections from the neurone (the dendrites) and the long projections (the axons). Small pores or holes in the cell membrane operate as gates, and when a neurone is activated, these gates open, allowing sodium ions to flood in. This generates an electrical pulse, called an *action potential*. This pulse activates the neurone and enables it to communicate with other neurones. Not all nerve cells are in your brain, and this electrical pulse is the way nerve cells throughout your body transmit information. This is why if you've ever had the misfortune to grab hold of an electric fence, your muscles contract very fast as the electricity stimulates the peripheral nerves in your arms and hands (as well as hurting like hell).

But how do neurones talk to each other? This happens back in the synapses – the connections between neurones. Neurones are stimulated – made to create an action potential that passes along the cell – by chemicals called neurotransmitters. These neurotransmitter chemicals are produced in the little buttons at the ends of axons and dendrites; when the nerve cell is activated, these are released into the gap between the two buttons, called a synaptic cleft. The chemicals move across the gap to the other button, where they bump into receptors in the surface, and affect the activity in that next nerve cell. The neurotransmitter can be excitatory, meaning it causes an action potential to be discharged by the next cell, which will pass the signal across further synapses. Or it can be inhibitory, meaning it *prevents* an action potential from happening. In addition to the complexities made possible by the creation or the prevention of action potentials, there are many different neurotransmitters in the brain, and any one neurone will be part of one neurotransmitter system. Changes to the brain's neurotransmitter systems can change the ways in which the brain works.

Thus, the neurones form networks, with synapses forming connections between neurones. *Within* a neurone, information is transmitted using electrical charges. *Between* neurones, information is transmitted chemically.

These patterns and networks of connections are what enable our brains to work as systems for processing information – for being you. Any time you learn something, or get faster at performing something, these changes are facilitated by changes in the connections between neurones – existing connections are strengthened or weakened; new connections are made. This plasticity means that, on a larger scale, the networks of neurones can grow and change, and our brains can learn and adapt. Your brain is still your brain, built largely of the same neurones as it ever was, but it changes from minute to minute, day to day, year to year, largely because of this continual pattern of plasticity and development. And while this development is most striking over the first few years of your life, your brain can and will remodel its connections and its networks over your whole lifespan.

Just as the form of neurones, with their tree-like structure, mirrors other kinds of natural systems that branch and divide (plants, blood vessels, rivers) the structure of the brain itself mirrors aspects of other systems where there is value in having a large surface area that needs to be folded into a smallish space – lungs, for instance, or gills, or leaves on a tree. The neurones are not all mushed into one mass in your skull: they are organised into a number of discrete structures, which on the whole follow the structures of the neurones themselves – the cell bodies (where the cell nucleus lives) are grouped together into small groups or much larger sheets, and the axonal projections that connect them form information superhighways of connections, large and small, that link the neurones together.

I'll discuss the precise anatomy of the brain in following chapters, but we can get a rough idea of the major landmarks by distinguishing between the cortex, the cerebellum, and subcortical structures. When you look at a human brain, the large, folded and wrinkly surface is the cortex. Meaning 'bark' (as in a tree) the cortex is the outer surface of the cerebrum, and contains the cell bodies of many neurones, their axons forming thick tracts of connections beneath the cortex. As the axons look white, due to their slim fatty sheaths, and the dense layer of cell bodies looks greyish, the terms white matter and grey matter are used to distinguish between these two kinds of structures in the brain, though they are comprised of the same neurones.

The cortex is therefore folded up into multiple peaks and valleys – fitting as much cortex into the skull as possible, without needing to grow a huge head. Unfolded, the surface area of the cortex is around twelve centimetres by twelve centimetres – roughly the size of a tea towel. The cerebellum, or 'little brain' sits below the cortex, at the back of the head: like the cortex, it's formed of a layer of cell bodies, folded into a smaller space. The folds, however, are different in appearance to the cortex; finer, almost fern-like. Both the cortex and the cerebellum are formed of two hemispheres – left and right. Beneath the cortex are a range of smaller nuclei of grey matter (nerve cell bodies), connected to the cortex and the rest of the rain by the white matter tracts. These are arranged symmetrically around the brain stem, mirroring the left/right organisation of the cortex. The brain stem – which sits at the top of the spinal column – is a structure that is critical to information getting into and out of the brain. The spinal column, also part of the central nervous system, feeds information into the brain stem about the body, and it's the conduit for the nerves with which we control the body. Another set of nerves – the cranial nerves – bring information into the brain stem about hearing, vision and taste, and control breathing, facial movements and articulation.

The scale of all this remains humbling to me. The

existence of 86 billion neurones, let alone all the ways in which they can form connections with each other, is hard to comprehend. I do a lot of brain imaging, and the smallest unit of brain that we work with (called a voxel – like the pixels on a TV screen, only 3D), contains approximately 630,000 neurones, and many more synapses. The neurones are all surrounded by a dense matrix of around 85 billion supporting cells – such as glial cells, which ensure that the neurones get enough nutrients, and clean up cellular waste – as well as a complex network of blood vessels. Your brain is only around 2 per cent of your body weight, but it uses around 20 per cent of the oxygen in your blood – so it needs a good blood supply to be able to work.

All of this is packed into a structure that can fit neatly into your skull, and almost all of those neurones were there when you were born. They are why you have a continuous sense of being you, over your whole life. When you look at a picture of you as a baby, you're looking at a body that contains a brain that is at once the same as yours – it's built by the same neurones – and also totally different, because human brains grow very rapidly in the first few years of life, and continue to be shaped by our experiences for the rest of our lives. Imagine: anything that you have learned while reading this first chapter has just changed your brain again.

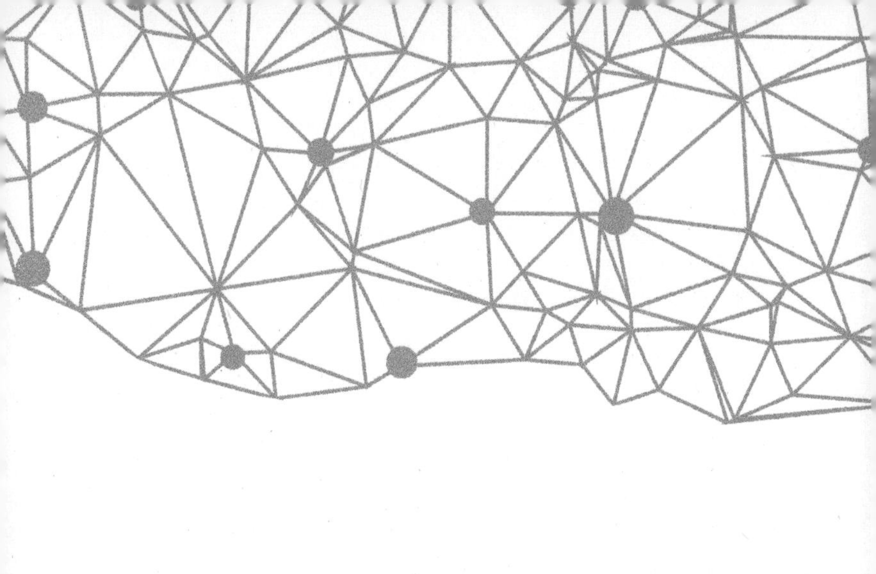

2

How Does Nature Build a Brain? Brains and Evolution

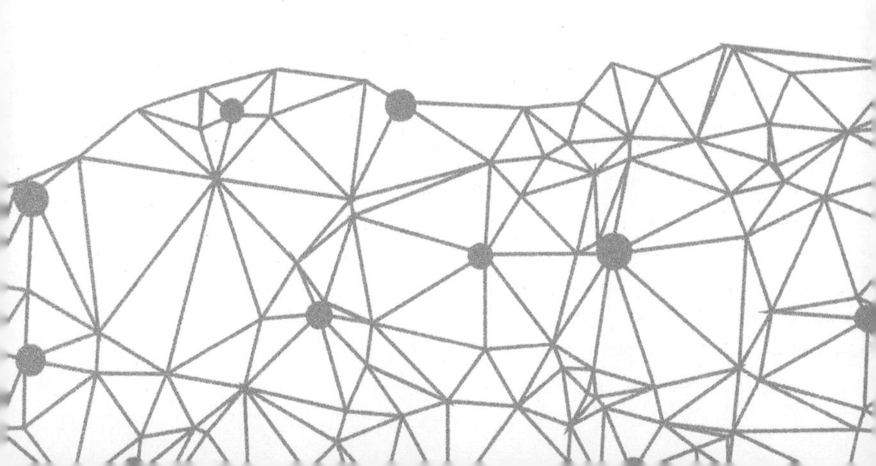

Why do you need a brain? Lots of lifeforms do not need a brain. Plants don't have brains or any kind of nervous system, and they sense and respond (albeit quite slowly) to their environment. Unicellular organisms – that is, tiny living creatures that consist of one cell (such as amoebae) – don't have brains, and they can not only respond to their environment, but can also hunt their prey (e.g. bacteria) by smelling them (detecting their chemical trails). They can also avoid predators, and have even been shown to be capable of learning. But as soon as you are bigger than one cell, and you need to move about and be capable of complex behaviours, then nervous systems and brains become critically important. Bigger bodies require more coordination, which calls for the evolution of more neural control. More complex behaviour also requires the evolution of more nueral control. And all that neural control requires a nervous system and (ultimately) a brain that's in overall control. Nematode worms are about one millimetre long and have 302 neurones. I am about five foot four inches, and I have (hopefully) about 86 billion neurones. Take that, nematode worms. But how did these nervous systems evolve?

To think about the evolution of brains (specifically in mammals) we need to think about how life on earth evolved, as the formation of the first unicellular creatures contained the genetic code that one day, many millennia later, made neurones possible. We also need to think about what kinds of life require a nervous system and possibly also a brain.

About 3.8 billion years ago, we find the first signs of life on earth, with the emergence of cells: the basic units of life. Cells share some basic characteristics: they are contained within a cell membrane, they can metabolise food to create energy, and they can (normally) replicate. These tiny atoms of life evolved in water, and water has a lot of advantages if you are a tiny bit of life – you won't dry out, useful things like food and oxygen can be dissolved in the water, and you can send signals by releasing your own chemicals into the water. However, a watery environment also presents cells with a problem – how do you maintain a constant environment inside your cell without lots of water molecules moving into you (via osmosis) and making you burst? Cells solve this problem by being wrapped in the cell membrane, which is made of lipid (fat) molecules.

The cell membrane itself offers some protection from the invading water molecules, but cells also evolved the ability to control their chemical composition by pumping certain common ions (formed when elements dissolve into water) from the surrounding water into the cell itself. For example, levels of potassium are normally around ten to thirty times higher inside a cell than outside. This is such a successful strategy that all living cells keep a constant internal environment via this mechanism. Incredibly, the genes that code for these gates predate all fossils that have been found – in other words, all life on earth has been made possible by this very early adaptation.

And the ability to control the movement of ions in and out of the cell is central when we think about the evolution of the ability to create biological electrical charges that lead to action potentials. In addition to altering the salinity of liquids, these ions carry electrical charges. If the cell pumps potassium ions into the cell and sodium ions out of the cell, we create an electrical difference across the membrane, as sodium and potassium ions carry different electrical charges. This difference across the membrane can be used to generate an action potential – the electrical pulse that neurones use to transmit a signal. When the electrical difference is discharged. In other words, evolution has led from cell membrane mechanisms that are essential to the origins of unicellular life, to the possibility of using these cell membrane mechanisms to generate biological electricity.

We also saw in the previous chapter that, while the transmission of information *within* a neurone is electrical, the communication *between* neurones is chemical. When in evolution do we start to see neurotransmitters appear? Again, we can find the origins of these in unicellular organisms. For example, some tiny unicellular organisms called choanoflagellates can live in large colonies, forming a carpet across rocks. They use chemicals to communicate with each other, and to coordinate their food-catching actions. This has been identified as a very important precursor to neurotransmitters, as it is a demonstration of cells using chemical signalling to coordinate the actions of a large number of cells.

Incredibly, by studying the ways that genes code for gates in cell membranes, the construction of synapses, and the production of neurotransmitters, scientists have found that these genetic codes were in place in early unicellular life, long before they could have any utility for nerve cells. We normally think of evolution as being reliant on genetic variation or genetic changes, but here we seem to see evolution as a result of repurposing existing genetic codes. In other words, the evolution of brains did not co-occur with entirely new genetic sequences.

So, 3.8 billion years ago, we see the emergence of tiny creatures that consist of one cell, and which hold the potential for brains. But these are not yet brains! Around 900 million years ago, we start to find in the fossil record the first signs of multicellular life, with structures like sponges, which are organised like a community of choanoflagellates, but which live in one colony with a body. Sea sponges are entirely sessile, and do not have a nervous system, but can and do react to their environment by releasing chemicals that flow through the sponge and cause it to change how it moves in the water. However, this kind of signalling can be diffuse, slow and hard to control.

As soon as animals start to consist of more than one cell, we start to see a nervous system emerging. This is seen in the appearance of special cells that can be used to send signals around the body. We have already met these in the first chapter – these are neurones, and they are effective at sending signals because they can be long and thin, and because they can form connections with each other, so that behaviour can be shaped and modified through learning.

The best guess we have is that neurones first evolved from epithelial cells – cells that form surfaces in the body (like your skin, or the lining of your mouth). The evidence for this comes from jellyfish, who can generate action potentials in their epithelium, and because in the development of embryos (for the vast majority of animals) the epithelial structures and the neurones both develop from the same part of the embryo.

The first creatures with neurones probably still did not have brains. Comb jellies evolved 730 million years ago, and jellyfish 680 million years ago. Jellyfish have neurones structured into nerve nets, which are organised into different systems – for example, some nets are coordinating sensory information about light and gravity, and others are coordinating movement. Indeed, box jellyfish have even more complex neural nets, and are effective hunters. However, they do not have a central brain: their neural nets are structured, and box jellyfish have some conglomerations of neuronal cell bodies, but there is no central complex of neuronal cell bodies that starts to resemble a brain.

There is disagreement about whether neurones evolved once or twice – once, in a precursor to both jellyfish and comb jellies: or twice, once in comb jellies, and separately in jellyfish (and thence to all animals).

Around 630 million years ago, there came the first appearance of bilateral symmetry – bodies arranged around a central axis, with the two sides of the body forming identical halves. This order almost certainly started with something that looked pretty worm-like, but this order ultimately led to almost all multicellular animal life on earth – including humans. This bilateral symmetry meant that these creatures had a head, and this is where brains started to develop.

Around 540 million years ago, there came the Cambrian explosion, when the sea started to fill up (according to the fossil record) with a dizzying variety of animals, many of them heavily armoured – which suggests that predation was extremely common. The examples of tracks made by animals at this time change from simple and linear to complex, suggesting that there were marked differences in behavioural patterns. All of this suggests that nervous systems were becoming much more complex.

One creature that appeared at around this time, *Pikaia gracilens*, is where we start to find evidence of the first 'brain'-like structure. *Pikaia* was short (about one and a half inches long), with a flattened body and a tail fin – it probably swam like an eel. Although it had no distinct head, it probably had a very basic brain. *Pikaia gracilens* is now extinct, but its modern-day direct relative, *Amphioxus*, has a simple brain with just 20,000 neurones. Studies of these brains indicate distinct structural complexity, with different types of neuronal structures. *Amphioxus* looks like a flattened worm – not very impressive, perhaps – but it may represent a starting point for the extraordinary world of brains.

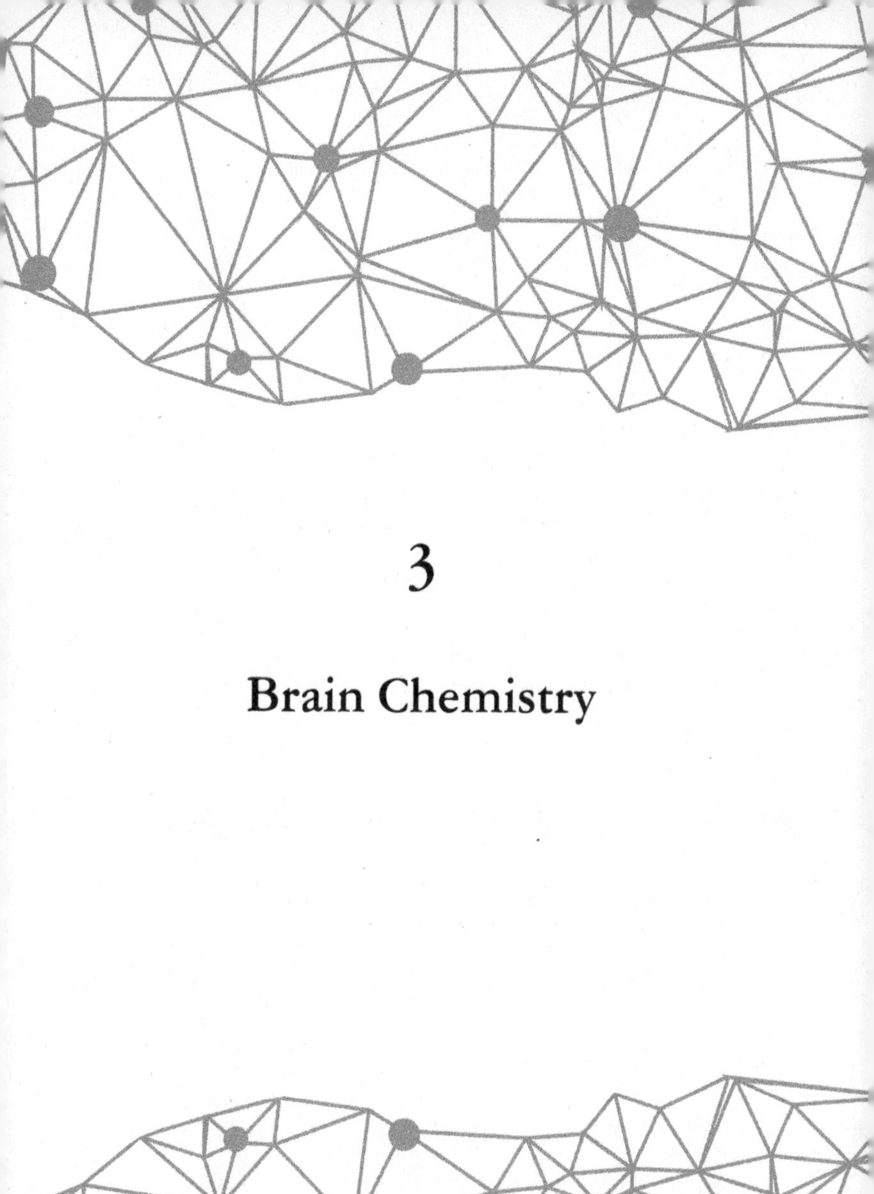

3

Brain Chemistry

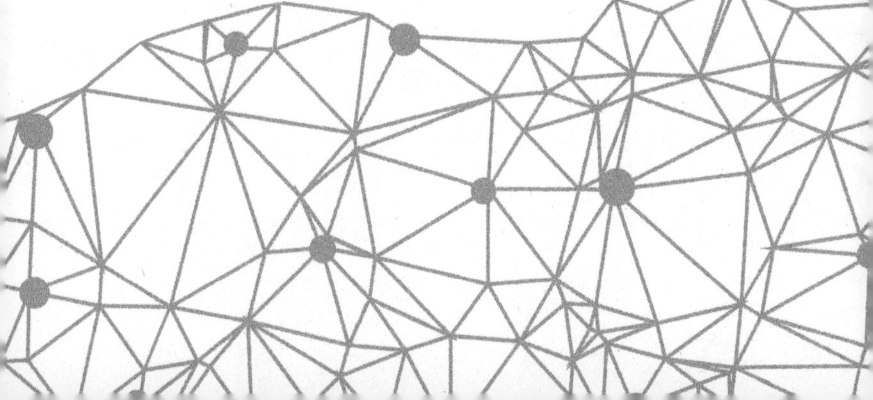

It is easy to look at the structure of the human brain, those 86 billion neurones, and feel astounded by the scale of the complexity that we encounter. But there is another dimension to this complexity – brain chemistry. While the signals that are propagated along any one neurone are electrical, the communications between this neurone and another neurone are mediated by chemicals called neurotransmitters. These are critically important in any nervous system.

You will remember that when a neurone fires, this causes chemicals to be released across the synaptic cleft – where neurones form connections with each other. When the chemicals are detected by receptors, this triggers that neurone to either fire – to send an electrical impulse – or to reduce activity. Thus, neurones can either activate or suppress other nerve cells. There are many different neurotransmitters in the human brain and body, and they are associated with a range of different functions. The neurotransmitters are synthesised within the neurones themselves.

Some neurotransmitters have very general effects on the brain. Two very important and widespread neurotransmitters in the brain are associated with excitatory and inhibitory effects. Glutamate is the principal excitatory neurotransmitter: this means that it makes neurones fire more. When neurones change their firing patterns and connectivity, glutamate is often involved, which has led to it being linked to the ways that information can be stored in the brain. In contrast, GABA (Gamma-aminobutyric acid) is a very important and widespread inhibitory neurotransmitter, reducing the activity of neurones.

Glutamate seems to play an important role in different kinds of degenerative diseases: glutamate excitotoxicity is a complex process by which excess glutamate causes the death of neurones. This has been suggested to be an important factor in Alzheimer's and Parkinson's disease. GABA forms the basis for drugs that try to control epilepsy, a brain disorder associated with abnormal patterns of electrical activity in the brain. GABA seems to control the excitatory state of the whole brain via these inhibitory effects; when this effect is unbalanced, seizures can occur.

Other neurotransmitters can have less general effects, and are associated with more specific brain regions. Dopamine is a neurotransmitter with a range of roles: it is involved in the initiation and control of voluntary movement and, separately, in the experience of reward. Dopamine is also important in the inhibition of milk production in mammals, in our moods, and in sleep and dreaming. Dopamine is an extremely important neurotransmitter and has been linked to attention, learning and working memory. Dopamine levels are affected by Parkinson's disease, and people with Parkinson's disease can have severe difficulty in initiating movement, and in processing positive and negative rewards: when off their medication, people with Parkinson's disease learn better when they are punished for wrong responses than when they are rewarded for correct responses. When they are on medication (which mimics the effects of dopamine) they show the opposite pattern, as do most neurotypical people.

Neurotransmitters are extremely important in our emotional processing. Serotonin is a neurotransmitter that modulates multiple neuropsychological processes and neural activity. One paper has suggested that it is hard to find to find a human behaviour that is not regulated by serotonin. It plays a role in our moods, our perceptual processes, our experiences of reward, rage and anger, our eating behaviours, our memories and our sexual behaviour. Interestingly, most of the serotonin found in our bodies is not in our brains. Serotonin receptors are found in most organs in the body, and serotonin plays an important role in the heart and blood vessels, the lungs, the digestive system and the bladder and bowels. Serotonin levels can be disturbed in some psychiatric conditions, such as anxiety and depression. For example, people with depression show lower levels of tryptophan, a molecule that the brain uses to synthesise serotonin. Many drugs that aim to treat depression target the levels of serotonin, and this can be successful for some people with depression (though not all).

Some neurotransmitters play an important role in specific emotions. Noradrenaline (also called Norepinephrine) is a neurotransmitter that is found in the brain and in the nervous system throughout the body. It is very important in the 'fight or flight' response to threat or danger, both in the brain and in the body: in the body, noradrenaline leads to stress-related changes, including pupil dilation, increased lung capacity, increased heart rate, increased blood pressure, and decreased activity in the intestines. In the brain, noradrenaline is associated with vigilance, arousal, attention, motivation, reward, and also learning and memory. Indeed, some studies show that increases in noradrenaline can work backwards on memories, making memories of events that made us stressed or frightened seem more salient and detailed.

Some quite common chemicals can play important roles as neurotransmitters. You may have taken antihistamines to cope with hay fever, and you may have noticed that some can make you dozy. This is because histamine is also a neurotransmitter that mediates the functions that keep the body in balance, a balance called homeostasis. Histamine is involved in making us feel more awake, in our experience of hunger and eating, and aspects of our motivational acts (different things that we feel we want to do).

Neurotransmitters are not only important in the brain – the function of our sensory systems and the activation of our muscles both depend on neurotransmitters. Acetylcholine is a neurotransmitter that plays a critical role in the motor system, mediating how lower motor neurones stimulate muscles. Acetylcholine is also important in perceptual systems, and in mechanisms of the ways that neurones can adapt and learn, such as memory systems and in attention networks.

There are some other important neurochemicals that affect neurones. Oxytocin and vasopressin are neuropeptides, important in the control of contractions during childbirth, and the control of urination respectively; they also play very important roles in mammal social behaviours. It is hard to study these in humans, but there is now evidence that increasing levels of oxytocin leads to increased attention to social information in humans. Vasopressin increases seem to be associated with improved cognitive function. Both oxytocin and vasopressin seem to have their brain effects mediated by the amygdala, a brain area that is very important for emotion processing, learning and processing social information.

Oestrogen and testosterone are often called sex

hormones, as their expression affects the development of sexual characteristics in the womb and during adolescence. Both also have effects on the function of neurones, though these effects are complex: many of the areas in the brain where testosterone is active are areas where it is changed into an oestrogen to have its effects. Oestrogen receptors are common in the brain, affecting aspects of mood and cognitive processes, as well as sexual behaviour and eating. Oestrogen also has neuroprotective effects in the brain, which may have important implications for the effects of the menopause on brain function, since oestrogen levels drop dramatically post-menopause.

A different kind of hormone is cortisol. Cortisol plays an important role in how your brain and body response to stress. It is also central to sleep cycles, and is associated with waking up in the morning (which may well be why waking up early can feel so horrible). In the brain, cortisol has effects that are mediated by the amygdala. Cortisol has slower-starting, longer-lasting effects on the body than noradrenaline, and living with high levels of cortisol is good for neither brain nor body in the long run.

Endorphins are a very important kind of chemical that can affect brain function. Endorphins are neuropeptides – molecules that affect both neural activity and also cell growth – and they are found in the brain and body. In the body, endorphins are associated with an inhibition of aspects of the pain pathways: this earns endorphins the title of the body's own painkillers. In the brain, endorphins suppress the activity of GABA (see page 32) which leads to an increase in the release of dopamine. Endorphin uptake by neurones is associated with exercise, singing, eating, dancing and laughing. There is also a theory that human social bonding is associated with endorphin uptake, and it's interesting that we particularly enjoy activities such as exercise, singing, eating, dancing and laughing when we get to share them with other people.

So you can see how immensely complex the brain's neurotransmitter systems are – and this is just a snapshot! It is common to see neurotransmitters described as performing simple, distinct roles – for example, to see dopamine described as a 'happy' neurotransmitter – and it's true that when you experience something really pleasingly rewarding, you do get a dopamine 'hit'. But dopamine is also critical to our ability to make voluntary movements. Serotonin, acetylcholine and noradrenaline are all linked to aspects of memory, but all differ in other ways. When you think of all the possible connections between the 86 billion neurones in your brain, and then imagine how these are all being affected by different neurotransmitter systems, you can see how unlikely it would be that any one neurotransmitter would be linked to just one function.

It is also worth considering how our brains respond to chemicals that we introduce them to. Any drug that changes brain function is psychoactive to some degree. Many drugs cannot get into our brains via our bloodstream because of the blood–brain barrier, which is a tight network of cells that surround the blood vessels in the brain and limit the molecules that can diffuse out from the blood plasma into the brain – in effect, this is a way of protecting the brain and its highly specialised neurones. It also means that not all medicines or drugs we consume will necessarily have their effects within the brain itself.

Alcohol is an extremely widespread drug: most human cultures have discovered it and used it in a wide variety of situations, from rituals to celebrations. Alcohol is a small molecule, so it can cross the blood–brain barrier, and it is very rapidly effective. Alcohol has a number of complex effects in the brain. It stimulates GABA receptors, which are part of an inhibitory network, and thereby dampens activity in the brain overall. This is why alcohol can be calming, but also makes people sleepy. It is also one of the dangers of alcohol: too much alcohol, and really important brain functions, like keeping you breathing, get suppressed. Alcohol also suppresses the release of glutamate, the excitatory neurotransmitter: this results in a slowdown of brain function, meaning that decisions and responses can become slower and less accurate. Both of these effects are associated with behaviour being more disinhibited.

Alcohol is associated with the release of dopamine, which can feel good, but that effect tends to reduce with continued use, by which point people have often started to drink just as a habit. Alcohol has a stimulant effect on the brain, causing a release of noradrenaline, and it is also associated with increased endorphin uptake. This increased endorphin uptake can feel good and acts as a painkiller, meaning, for example, that falling backwards into the bath while attempting to take off your coat (as I managed to do while celebrating the millennium), doesn't hurt until the next day. Alcohol is also associated with increases in cortisol release, which, as noted above, is a hormone associated with waking up and also with stress. This means that, despite the sedative effects of alcohol, sleep is often poorer and more disrupted after drinking it.

Caffeine is probably the most widely consumed psychoactive drug around the world, as it is found in tea and coffee and has a mild stimulant effect. This is due to caffeine's relationship with adenosine, a neurotransmitter that builds up throughout the day, and has an inhibitory effect on brain function. Adenosine plays an important role in the way that we start to feel tired at the end of the day, and want to sleep. Caffeine works by binding with the adenosine receptors and preventing the build-up of the inhibitory effects of adenosine, hence reducing sleepiness. Caffeine also seems to have a positive effect on attention, possibly through interactions that adenosine has with dopamine. Although caffeine is only a mild stimulant, it is highly addictive, and caffeine withdrawal is disproportionately unpleasant.

Anything that is psychoactive – from legal products like alcohol and caffeine, to illegal drugs like amphetamine, and medical drugs for psychological and psychiatric and neurological problems – will affect your brain chemistry, and these brain changes can be why they are popular or effective. It's also probably a good reason to think carefully about anything that we consume that can have effects on our brains.

4

How Do We Know About the World?

The only way that we know anything about the world is through our senses, and the ways that our brains process and interpret the information they provide. Everything we experience is just the brain's best guess about what is out there. There is a lot of information in the world, and organisms differ in the kinds of sensory information they can process. I will start this journey with a decidedly mammal-centric perspective on senses, though I will expand this when we talk about how brains in different animals vary.

It is very hard to discuss the brain, a three-dimensional structure of almost absurd complexity. I hope you will forgive me for resorting to a range of analogies. One quite useful analogy is to make a fist of your left or right hand – looking from the thumb side, this looks quite like one of your two hemispheres. Your folded fingers form the frontal lobe, your thumb is the temporal lobe, and back of your hand is the parietal lobe. If we enter the realms of fantasy and imagine that our hand ends at the junction with the wrist, this is where we find the occipital lobe.

All our senses rely on specific organs or receptors that are specialised for particular kinds of information, and which can turn that information into an electrical signal that can be sent to the brain for processing.

Depending on the kind of information being processed, this can make perceptual systems really quite different.

Taste

Taste and smell are called the chemical senses, because they rely on the detection of actual molecules striking our receptors. When you smell or taste something, you are processing a tiny bit of the thing itself, which is a slightly disconcerting thought. The chemical senses are the oldest senses, in an evolutionary sense – unicellular organisms can release chemicals to signal to others, and detect chemicals released by others. As unicellular organisms evolved in watery places, these chemicals diffuse through the liquid environment. And, of course, this remains the case for taste – the saliva in our mouths provides the liquid that makes taste possible.

Our taste system is pretty basic. We are sensitive to just five different characteristics of chemicals in taste: salt, sour, bitter, sweet and umami (savoury). These are based on two different kinds of receptors, which live on our tongues, in our taste buds. Salt and sour receptors are rather basic ion channels, effectively not much different from receptors that we find in sea sponges. Sweetness, bitterness and umami tastes are detected by more complex receptors, but they are still only processing a single aspect of the food and drink in our mouths. And if you are mad about chilli, you may be surprised to learn that we do not have receptors for the taste of chilli – chilli is an irritant, and the chilli 'heat' is detected by pain detectors in the mouth.

The detectors in taste buds send electrical information via the facial nerve (cranial nerve VII) and the glossopharyngeal nerve (cranial nerve IX). These nerves project up to the brain via a sequence of tiny relay stations in the brain stem and the thalamus, and then up to the gustatory cortex. The gustatory cortex, so named because it's the first point in the cortex that receives this taste information, is tucked away inside the insula and the frontal operculum. If you make your hand into a fist again, where the thumb is like the temporal lobe and, the insula sits on the inside of your thumb where it touches your forefinger, and the frontal operculum is under the first joint of the forefinger.

At this point, taste gets fully integrated with incoming smell information. What we actually experience as taste is an integration of the output from the taste receptors, and what we are smelling as we eat and drink. If you have a cold, you may notice that food tastes dull, and that's because a blocked nose stops you from being able to use smell to enhance taste.

Smell

What we experience as smell is based on airborne transportation of molecules. As I've explained, chemical senses evolved in water, but smell became an incredibly effective tool when animals colonised the land, where molecules are carried by air. We have around 400 different kinds of smell receptors in the lining of our noses. These are cells with fine, hair-like projections, sensitive to certain molecules, and we have many more of them than we have taste receptors. When our smell receptors detect certain molecules, we experience smell, although any one smell will be made up of a wide range of different chemical elements. And we have many of each kind of receptor – hundreds of thousands of each. These receptors link to the olfactory nerve. In humans, this nerve projects straight through little holes in the skull into the cortex, bypassing the brain stem and the thalamus. The olfactory nerve projects directly into the olfactory bulb, which sits at the bottom of the front of your brain (under the fingernail of your forefinger in the fist-brain analogy). From here, it can project the information to the nearby gustatory cortex, so that our experience of taste incorporates both kinds of information. I sometimes wonder if this lack of detailed processing of smell in the brain (in humans) is the reason why we can be very interested in smell, yet simultaneously struggle to describe any one scent. I can have a go at saying why a flute sounds different to a piano, but I would have great difficulty saying how coffee smells different to a rose. This almost certainly reflects the reduced importance of smell to humans – other mammals live in a world of smell and have very large parts of their brains devoted to it. For humans and other primates, sound and vision have become more important than smell, and that's reflected in a great reduction in the size of the areas that process smell in the human brain, combined with a reduced vocabulary for smell. It is striking that when people train as tea tasters or wine experts, they learn to map taste and smell onto a much wider range of labels.

Vision

Vision is also a chemical sense, but a rather different one: our eyes detect light using chemical sensors that turn light energy into electrical impulses, which then send information to the brain via the optic nerve. There are two kinds of receptors, rods and cones, which line the retina at the back of your eye. Cones are better at detecting colour, and they are densely packed into the fovea, which we'll learn more about later. Rods are found towards the edge of the retina, and detect edges and movement.

The retina is a 'highly conserved' structure in evolution – it is found in the eyes of most vertebrates and some molluscs. Because of the ways that the retina develops in the embryo, it is considered to be part of the central nervous system (i.e. the brain) that has relocated to the eye. Information in the retina is encoded into the optic nerve in a way that preserves the spatial layout of the visual world around you. This is preserved all the way up to the cortex, with a couple of wrinkles: first, before the visual information gets to the thalamus relay station, it goes through the optic chiasm, located in front of the brain stem, where there is a crossing-over of some of the visual detail from each eye, meaning that all the information from the right-hand side of the world is sent to the left side of the brain and vice versa. When you see the world, your visual system is gluing the left- and right-hand sides of the world together.

Second, after the visual information is sent to the primary visual cortex, it is represented in a highly non-linear way – much more of your visual cortex is devoted to the fovea and the area around it, and much less to the surrounding visual field. The fovea, which is the bit of the eye that we look at things with, has the best resolution, while right out at the visual periphery – the edges of our vision – we have very poor resolution and don't even have good colour vision. Of course, this is not what our sense of vison feels like. As I sit at my desk, it seems to me that I have a rich and detailed visual experience. My visual world feels a bit like a cinema screen, and I can choose where I want to look within that. In reality, at any one time, the visual detail to which we have access is extremely limited, and when we look at something, we only have detailed visual information about the tiny area around where we are looking. Now, sitting here at my desk, it doesn't feel like the edges of my vision are poorly detailed, without form and leached of any colour, although at any one point in time, that's exactly what they are.

The main reason our visual world feels detailed and complete, and not like we're only seeing a tiny bit of detail surrounded by blurriness, is that we are moving our eyes all the time. Absolutely all the time. Two to three times a second. I don't mean the kind of eye movements where you find your eyes tracking something moving, like a car going past. Instead, I mean the jumps and leaps that your eyes make when you are simply looking at something. These periodic jumping movements are so constant that we don't even notice that we are doing it, but we are continually mapping out our visual environment.

Once visual information reaches the cortex, it is processed in very complex ways, with some anatomical pathways pulling out information about what the things around us are – categorising objects, recognising faces, decoding what words mean. These pathways run forward along the bottom of the temporal lobe. Other visual pathways run up into the parietal lobe, and coordinate the incoming visual information with the movement of our bodies – for example, to guide ourselves walking down steps, or to help us put a key in a lock. And, of course, we use visual information to guide the movements of our own eyes. I will cover these links in the brain between perception and action in chapter 5.

Hearing

What we experience as sound is a particular class of vibrations – the same vibrations we can feel as the rumble under our feet when a tube train passes by. The sounds we hear are mostly vibrations of the air molecules around us. The job of the auditory system is to turn these vibrations into electrical discharges. The human ear (like all mammalian ears) turns vibrations into sounds via a sequence of steps. When something makes a sound, this vibrates the surrounding air molecules, and when these vibrations reach our ears, they move our eardrum, a tight membrane that separates our outer and middle ears. Behind the eardrum are three tiny bones, forming a small bridge that transmits the vibrations across our inner ear to the cochlea. The cochlea is a curled structure containing fluid, and the vibrations transmitted from the tiny bones moves this liquid. These movements then push cells called hair cells up and down. Hair cells have long, hair-like projections, and when a hair cell moves, these projections bend. This bending triggers an electrical signal to the auditory nerve. It's quite extraordinary to realise how many physically moving parts your ears contain, just to get sounds to your brain – and this explains why it is so easy to damage hearing by exposure to loud sounds – these moving parts can and do break and wear out.

We saw how vision preserves the spatial array of visual information. Hearing preserves 'tonotopy' – the frequencies of sounds, from low to high. This is captured first at the cochlea, and is maintained all the way up to the auditory cortex, but sound has a complex journey to get to this cortical destination. Auditory information is highly processed in a series of nuclei in the brain stem before reaching the relay station of the thalamus, and being directed to the auditory cortex. These nuclei seem to be essential for us to hear, for example, where a sound is coming from – as we only have two ears, we have to work out the spatial location of sounds by comparing the input to both our ears. And that is not all – these nuclei also start to group sounds together from different sources – essential when you remember that we rarely hear sounds against a silent backdrop.

The primary auditory cortex sits on the top of the temporal lobe – using our fist-brain analogy, the primary auditory cortex is located just in front of the crease at the back of the thumb, and is hidden inside the fold between the thumb and the edge of the hand. Just as in the visual system, once sound is represented in primary auditory fields, it is directed off to different anatomical pathways. One runs forward down the temporal lobe, and is associated with the recognition of different sounds – what different words are, for example, or different environmental noises (such as toilets flushing or doors slamming). This network seems to be particularly important for engaging with the wider language system when we understand spoken language.

Another pathway runs backwards, up towards the parietal lobe, and this is important for both processing where sounds are in space, and for guiding our own sound-producing actions, such as talking, singing or playing musical instruments.

Somatosensation (touch)

Touch receptors can be found within our skin and our mucous membranes. Touch itself is made up of several different kinds of sensation which each have distinct receptors.

The first of these receptors is pressure. We sense pressure when something presses on our skin, and this distorts cutaneous mechanoreceptors. A bit like the hair cells, these need to be moved to send a signal to the brain. The sensation of pressure is very important – it can tell me that I have hit a key on my computer and that I am still sitting on my computer stool. There are different kinds of mechanoreceptors, which vary depending on what kind of skin they are on. Generally, humans have skin that is classified somewhat bluntly as 'hairy skin' (such as the skin on our arms) or glabrous 'non-hairy' skin (found on the palms and fingers of the hands, the soles of the feet, and the lips).

The first of these receptors are Merkel cells, which have a very precise degree of spatial detail – they are particularly found in the fingertips and are especially good at processing edges and shapes (like when you're trying to find your keys in your pocket). They can differentiate between textures when you run your fingers over surfaces.

Tactile corpuscles are receptors found in the finger-tips, and lips and are very sensitive to light touch. These respond much more quickly than Merkel cells. You will notice that non-hairy skin is found on parts of the body that we use to actively interrogate our worlds, such as our hands and lips.

Found through the rest of the (hairy) skin, Pacinian corpuscles are receptors that detect broad textural differences, such as rough or smooth, and are responsive to touch in a very unprecise way: they do not resolve space very well. If you close your eyes and ask a friend to touch your back, you will find it very difficult to work out how many fingers they are touching you with: if you do this on your lips, you will find it very easy.

All these different kinds of sensation from our skin (both hairy and non-hairy), including those from pain and temperature receptors, are directed up the spinal cord, where they are processed in the brain stem and thalamus relay station, and then sent to the primary somatosensory cortex. Looking at our brain-fist, this is a stripe of grey matter in the parietal cortex, running from the fold of the forefinger up to the knuckle. The representation that is preserved here is a map of the human body, meaning sensations are being mapped onto where they were felt in the body. However, the map doesn't follow the size of the different body parts. Instead, the body parts where we have more detailed touch sensations are represented as larger, so there is a lot of space on this map for the face, lips, tongue and hands, and much less space devoted to the back or the legs.

Proprioception

You often hear about us having five senses, although we humans have at least eight: they may seem less dramatic, but they are extremely important. Sometimes classified alongside somatosensation, proprioception is our sense of where our body parts are in space. If you close your eyes and touch your nose with your index finger, it is proprioception that helps your brain guide your finger to the right place. Proprioception relies on receptors in our joints and tendons, and problems with proprioception can make 'easy' tasks like getting dressed incredibly hard.

Balance

Our sense of balance give us the ability to move around in the world, by allowing us to coordinate our sense of moving with our own actions. The vestibular canals in our inner ears interpret our bodily movements by being sensitive to gravity and linear acceleration. They are filled with fluid, with tiny stones attached to hair-like projections (very similar to aspects of hair cells). When we start to move, or change direction or change speed, the stones move in the fluid and bend the hairs: this then sends a signal to the brain along the vestibular nerve. There is still much to learn about this pathway, but it probably projects to the insular cortex. This is adjacent to the auditory cortex, but is on the inside of the temporal lobe, while the auditory areas are on the outside. One critical role for this vestibular pathway is to interact with the control of eye movements – when there is a mismatch between these two networks, we can start to feel extremely unwell (e.g. travel sickness).

Interoception

This is the sense that relays information about many different tissue types and organs from the body to the brain. This can include heart rate, respiration (breathing), the genitourinary system, gastrointestinal system and thermoregulatory system. Interestingly, the information from these different networks does not track to one single pathway to the brain, but has inputs to many different brain regions. This is unlike all the other sensory systems that we have addressed, and points to a wide range of different ways that the brain can engage with interoceptive information.

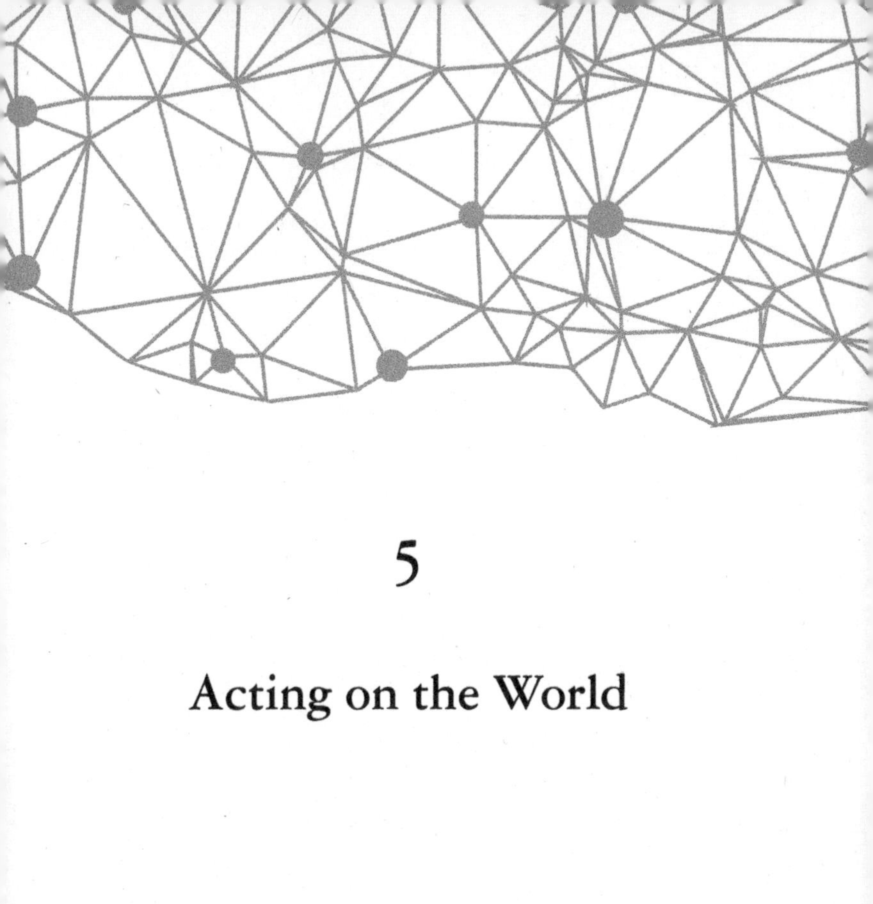

5

Acting on the World

When we interact with the world, we make movements – we move our hands, legs and mouths. We raise our eyebrows, we wiggle our toes. All this movement is made possible by a kind of hierarchical network in our brains that specifically has control over the actions we make with our skeletal muscles (as opposed to the muscles in our hearts or our guts). No matter what the kind of motor control involved (and there are a few different kinds), the end point is always stimulation of the muscles via the neuromuscular junction. There are two kinds of motor neurones: lower motor neurones, which project to the muscles from the brain stem and the spinal cord, and upper motor neurones, which connect from the rest of the brain to the cranial nerves and the lower motor neurones. It's a crazy thought, but when you wiggle your toes, there are only two neurones between your brain and your toes that cause that to happen.

We can start at the level of the lower motor neurones that stimulate our muscles to move. These work like sense organs in reverse: the electrical impulses that pass down these neurones stimulate the synapses at the neuromuscular junctions, causing the muscle to move. But how does this message get to a muscle to make it move?

There are four broad kinds of ways that our brains and nervous systems control movement. The first, and simplest, are involuntary reflexes. For example, when a doctor taps the front of your knee and your leg swings up, that is an involuntary reflex, mediated solely by the sensory neurones, the lower motor neurones and the spinal cord. The tap stretches the ligament at the front of the knee and this sends a signal to the cell bodies in the spinal cord, which detect the stretch, and a message is then sent to the muscles around the knee to move to compensate for the perceived stretch. There is no involvement of the upper motor neurones or brain at all. Some involuntary reflexes are routed via the lower motor neurones in the brain stem – for example, the responses of the pupils, the dark apertures at the front of your eyes, which shrink in size when you shine a bright light on them, and the gag reflex, which is triggered by tickling the soft palate. Highly over-learned behaviours, such as how to walk, are also mediated by these spinal and brain-stem reflexes.

The second motor system consists of the upper motor neurones, whose cell bodies lie in the cortex or the brain stem. They project down to lower motor neurones and thus to muscles. Upper motor neurones in the brain stem are important for how we control our posture, and how we orientate ourselves to our environment based on information from the eyes, ears, vestibular (balance) organs and our interoceptive states. When you hear a loud sound and you turn to look at it, the upper motor neurones in your brain stem coordinate that action.

In contrast, upper motor neurones in the cortex are associated with more complex behaviour, such as controlling voluntary movements and complicated sequences of movements. In the primary motor cortex, we find the cell bodies of upper motor neurones arranged in a map of the body. This lies directly in front of the somatosensory map of the body (see page 54), running down from the knuckle of your brain-fist, towards your palm. Again, the map of the body in the primary motor cortex is not linear – the finer the control that we have over a body part, the larger a space in the primary motor cortex is devoted to that body part. So, we have a lot of this map devoted to the face, lower lip, tongue and hands, and much less to the toes, the knees or the shoulders.

The premotor cortex (just in front of primary motor cortex) also contains neurones that directly influence the upper motor neurones and both these regions are critically important for coordinating complex movements. Stimulate a region of the primary motor cortex and you get movements of single muscle groups: a finger might twitch, or your whole leg. If you stimulate the premotor cortex, you can see more complex sequences of movements, or even a complete inability to move. This can be combined with a lack of awareness that planned movements are not actually happening, which suggests that these premotor fields are important for both planning and controlling complex sequences of movement, and maybe for elements of awareness.

There are two other important systems involved in motor control. These talk directly to upper motor neurones and can modulate how they function. The third motor system involves the cerebellum (literally, little brain), which sits at the back of the brain, under the visual cortex. The cerebellum outputs to the upper motor neurones, detecting errors in actions and automatically correcting them. The role it plays in error correction makes the cerebellum very important in motor learning.

The fourth motor system comprises the basal ganglia, the final critical element of motor control. These are small nuclei of grey matter that form a ring around the brain stem and the thalamus, that are very important in the initiation of voluntary actions in the upper motor neurones, and in the suppression of actions that are inappropriate. For example – if you pick up an unexpectedly hot saucepan, your reaction would be to drop it, but if it would be even more dangerous to drop it (e.g. it's full of boiling water), you can override this with your upper motor neurones and your basal ganglia.

When we act on the world, we use sensory information to guide our actions. If you have a dental anaesthetic and half of your mouth is numb, it is very hard to speak, because you use the sensations inside your mouth to help guide your articulations. We can see the importance of this sensory information in the motor system, as sensory parts of the brain (e.g. the visual, auditory, somatosensory cortexes) feed into the primary motor cortex.

One of the most interesting aspects of action control in primates is that they have both a greater amount of voluntary control over actions, and also more complexity in the control of these actions. The greatest complexity of these systems is found in humans. It is interesting to speculate about the ways that the evolution of structures like hands correlates with the evolution of the neural control of these structures. When I go to London Zoo to look at the squirrel monkeys, our hands don't look all that different (apart from the scale), but they cannot move their fingers independently. This works fine for important uses of hands by primates, such as holding a branch with the whole hand. By the time we get to chimpanzees, we see the use of precision grip, the fine control of the forefinger and thumb to manipulate objects. However, only humans are able to use precision grips with force, meaning that we can use this grip to operate needles, pens and cutting tools. Part of this is anatomical – human thumbs are longer than the thumbs of other primates – but it also reflects changes in the neural control of our hands. Many of the extraordinary achievements of humans come from our incredible bodies and our brains' fine control of them.

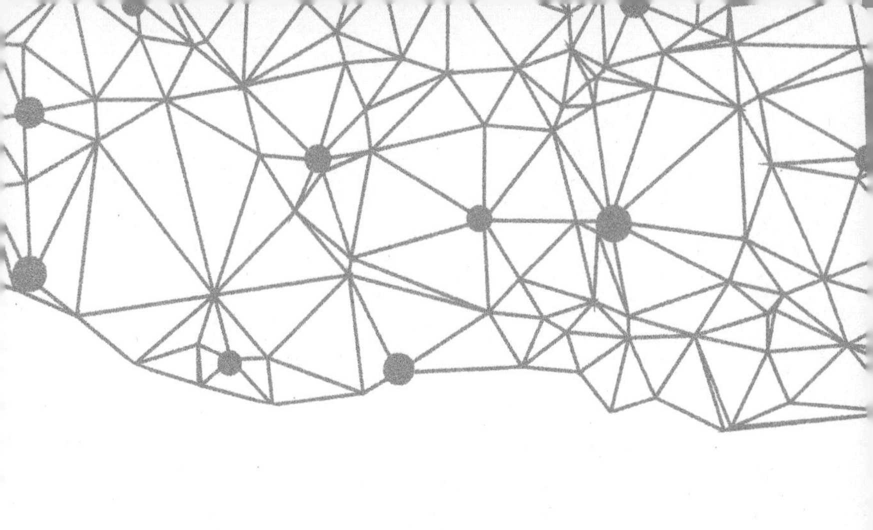

6

The Wider World of the Human Brain

We have seen how information gets into the brain, and how we can use our brains to control the world: in this chapter, we will consider all the bits 'in between' – the many different systems humans have that intervene between perception and action and contribute to the enormous flexibility of the human brain.

Association cortex

In addition to all the sensory and motor cortical fields I have discussed so far, our brains contain areas of association cortex. This refers to parts of the cortex that do not connect directly with incoming sensory information, or outgoing motor control, but seem to bridge between perception and action. Compared to other primates, the association cortex in humans is greatly expanded, suggesting that one of the key features of human brains is that we have more computational power intervening between perception and action.

The association cortex covers most of the cortexes in humans, apart from the primary sensory, primary motor and premotor cortexes. Association cortical areas take information from the sensory and motor cortexes, and are integrated and expanded upon to enable more complex thoughts and behaviours. Areas of association cortex projects to the hippocampi, the basal ganglia and cerebellum, the thalamus, and to other areas of the association cortex. The largest areas of association cortex in humans are found in the temporal lobes, the parietal lobe and the frontal lobe.

Association cortex – temporal lobe

The temporal lobe seems to be very important for recognising things in the world. If we go back to our 'make a fist' model of the brain, the temporal lobe is the thumb. Auditory information is coming into the superior (or top of the) temporal lobe, and visual information is coming into the inferior (or bottom of the) temporal lobe, from the occipital lobe at the back of the brain. Along the top of the temporal lobe, spoken words are being recognised, and if we look at the visual parts of the temporal lobe, running along the bottom of the brain, we find visual areas that process various kinds of visual information, distinguishing faces differently from objects, and written words differently from faces.

Very generally, one role of the temporal lobe is refining and coordinating visual and auditory information, meaning that by the time we get to the front of the temporal lobe, we find brain areas that respond to meaningful information in the same way, whether that information gets in via sight or sound (for example, as written or heard language). Semantic dementia is a brain disorder that initially affects the front ends of the temporal lobes, and the first symptom is often not understanding what words mean. Patients with this horrible disease can still hear words and can still use objects correctly – they might not understand what 'marmalade' means, but they know to spread it on their toast rather than rub it in their hair. As the disease progresses, and more of the temporal lobe is affected, they start to make more errors when they speak, and start to have difficulties repeating words.

There are important hemispheric differences in temporal lobe responses – language processes, such as reading words or listening to speech, are largely found in the left temporal lobe, while the right temporal lobe is very important in processing non-verbal information, such as faces, eye gaze, voices and music. People with semantic dementia typically have damage that starts at the front of the left temporal lobe; people with a similar pattern of damage that starts in the right temporal lobe tend to initially present with personality changes, possibly due to their difficulty in understanding the social meanings of voices and faces.

Association cortex – the frontal lobes

Looking at the 'make a fist' brain, the frontal lobes sit at the whole front of the brain, in front of the knuckle joint with the forefinger. This lobe starts with the strip of primary motor cortex; then, moving forward, we have the premotor cortex, which contains specialised fields for the control of the hands, the eyes and the breathing and articulation we use for speaking. There are important asymmetries here: the left primary motor fields control the right side of the body and vice versa. Therefore damage to one side of primary motor cortex can give people significant motor problems on the opposite side of the body. The language asymmetry seen in the temporal lobes is also seen here – the premotor fields associated with language production (talking and writing) are most often found on the left side of the brain.

Moving forward towards the front of the brain, we find the frontal association cortices: these seem to be especially important in integrating information and planning responses, and in inhibiting inappropriate responses. Patients with damage here – for example, due to being thrown forward during a traffic collision – can often make a good recovery, and show normal IQ and language use; however, when they return home or go back to work, they can show unusual patterns of decision-making or behave inappropriately. They may make disastrous financial decisions, or be more disinhibited in their language. Our big frontal lobes are also implicated in what is called 'novelty-seeking behaviour' – the tremendous curiosity of humans (and other primates) for the new and exciting.

Association cortex – the parietal lobes

The parietal lobes run back from the frontal lobes towards the back of the brain – they contain the somatosensory cortex, and link into visual and auditory areas of the brain. The parietal lobes seem to be very important in representing our bodies in our environments, and in linking sensation and action; they also seem to play a very important role in how we pay attention to the world around us. Damage to the right parietal cortex, for example, as a result of a stroke, can result in 'left neglect', where people do not pay any attention to things on their left side, although they can see and hear them (that is, the information *is* getting into the brain). This is a quite disorientating problem – people with left neglect may not respond to someone who speaks to them from their left-hand side, or may not eat food on the left side of their plate. When asked to copy a picture of a face, they may only draw the right-hand side of the face. Intriguingly, part of the problem may be because the left and the right parietal lobes both have attentional systems. The one on the left directs attention to the right side of our surroundings, and is very focused, for example, on things that are being done with the dominant hand. The attentional system on the right seems to be much more diffusely orientated to everything around us. When this right hemisphere system is damaged, we see the effect of the very focused left hemisphere attentional system, always pulling attention over to the right-hand side. So, 'left neglect' is almost more like 'right over-attention'.

The cingulate

The cingulate runs along the middle of the brain, above the corpus callosum, which is a giant white-matter tract that is the main superhighway for the two hemispheres to talk to each other. The cingulate is involved in a lot of processes. The front end of the cingulate, the anterior cingulate, sits just behind the frontal lobes, and is important in attentionally demanding tasks, such as error detection, or when you're trying to ignore distracting information. The anterior cingulate is also very important in the generation of emotional sounds – if you drop a frying pan on your foot and shout aloud in pain, the anterior cingulate cortex has a hand in this.

The insula

If we go back to our brain-fist model, the insula runs along where the inside of the thumb touches the palm and forefinger. Like the cingulate, the insula is involved in a wide range of brain processes, but it seems to be especially important in processing interoception, pain and a sense of self: the left anterior insula is also critical in the control of articulation when we speak aloud.

The hippocampus

Looking within the temporal lobe, formed by the thumb in the brain-fist model, we find some extremely important brain structures. The hippocampus is a structure that looks very slightly like a seahorse (hence its name), and it runs down the middle of each temporal lobe. The hippocampus has critically important roles in spatial processing and in memory formation. Research shows that there are cells in the hippocampus that map out where we are in a particular space (place cells), and these interact with cells in the surrounding entorhinal cortex that map out how we move through space (grid cells). This forms the neural basis for mammals to navigate their worlds. London taxi drivers who have completed the 'knowledge' (an intensive, years-long training in navigating London's streets and landmarks) show larger areas in the hippocampus than people who start the knowledge but who do not finish it, or London bus drivers who drive to preordained routes. The hippocampus is also central to our ability to form new memories: throughout the day, we encode memories into the hippocampus; then, when we sleep, our brains integrate this new information into brain regions surrounding the hippocampus. There is a strong suggestion that what we experience as dreams, which happen during specific stages of sleep, are glimpses of this process and of our brains' attempts to make sense of this. This is also why we learn new information better if we have had a chance to sleep on it, as what we have learned is being integrated into the rest of our existing knowledge and memories while we sleep. The hippocampi are often focal targets of Alzheimer's disease, sadly, which may be why people's abilities to form new memories can be compromised alongside increasing spatial disorientation.

The amygdala

The amygdala sits at the front of the hippocampus in both temporal lobes. It is extremely important in emotion processing, the formation of emotional memories, and face perception – it is not unusual for patients with amygdala damage to have problems recognising people from their faces (known as face-blindness). The amygdala also contains very important processes that mediate our ability to learn from experience, and to respond rapidly to threatening stimuli. Indeed, patients with damage to the amygdala can tell you what would make them scared, but may struggle to recognise a frightened face.

The hypothalamus

The hypothalamus is a small structure containing many tiny nuclei that sits just below the thalamus. It controls body temperature, our sleep/wake cycle, the release of oxytocin and vasopressin, the release of human growth hormone, the control of aspects of feeding and the feeling of fullness, and the control of heart rate and blood pressure. The hypothalamus is thus critical to the brain and body being kept in balance.

The brain stem

When I was mapping out perception and action in the brain, I discussed how the brain stem, at the top of the spine, has many nuclei feeding information into and out of the brain. The brain stem also contains structures that are critical to our being conscious, to breathing control, and to the control of vomiting. As the brain floats inside our skulls, which form an enclosed space, anything that raises the pressure inside the skull, for example, due to bleeding from a ruptured artery or brain swelling, can start to squash and distort the brain, and it is often the brain stem being forced downwards by the increasing pressure that produces critical symptoms. This is why vomiting and changes in consciousness should be monitored following a head injury, as these could be signs that the brain stem is being compressed.

We have seen over these past three chapters how extremely complex the human brain is in terms of its anatomy. But big brains can be risky, and simple brains can be very effective. Some colleagues of mine at a conference in Florida had a day off and went on a bicycle ride. They were very interested when they came across a dead alligator, and they and their big, curious brains got off their bikes and went to have a look at it. They got closer and closer to the dead alligator; then they threw stones at the dead alligator, getting closer still. At this point, the very-much-not-dead alligator and its tiny brain, having let them get close enough, jumped up and chased them quite a long way, and they were too scared to get back to their bicycles. Bigger is not always better.

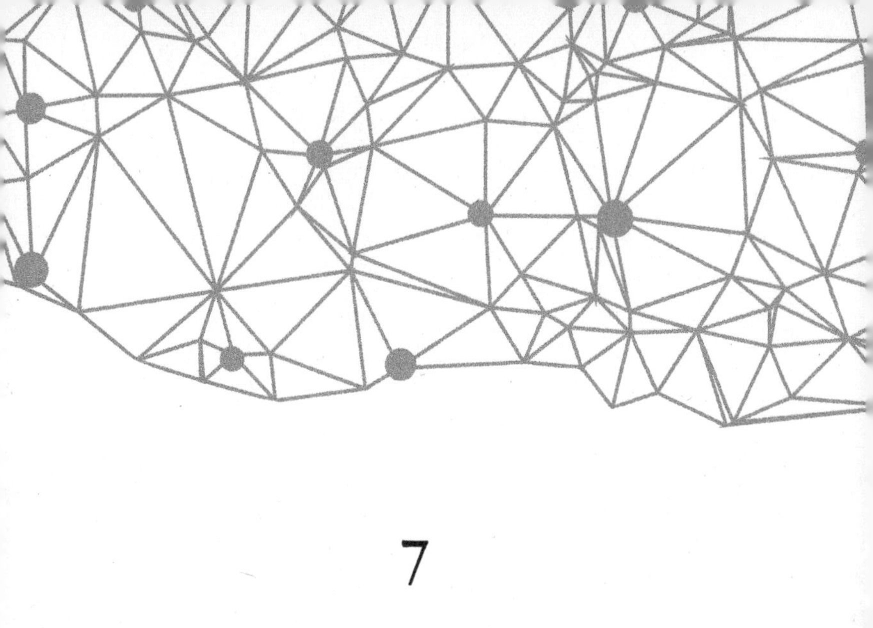

7

Different Bodies, Different Brains

This book has taken an unashamedly human-centric view of the brain. But evolution has also shaped the brains and bodies of all animals, and in this chapter I would like to spend a little time looking at a range of different animals and their brains, and explore how their brains relate to their bodies, the ways they sense the world, and their behaviour.

Alligators

As I ended the last chapter with an alligator outwitting a bunch of neuroscientists, let's start with their brains. There is an alligator skull at London Zoo: a slab of thick bone, about forty centimetres long, with holes for the eyes and the nostrils, some terrifying teeth, and, right at the back, a tiny hole into which I can just fit the top joint of my forefinger. And that space would have once contained its whole brain. However, with this relatively tiny brain, alligators can still achieve a lot – they are apex predators who sit at the top of the food chain. They are very vocal communicators, using a range of different sounds to signal to other alligators. They can even be trained to recognise their names (though I'm not convinced they would make great pets). So they are doing a lot with these tiny brains!

The alligator's small brain has some more familiar structures – there are elements of the brain stem and a thalamus, a small cerebellum, a very small cerebral cortex, and a relatively large olfactory area. This is consistent with their excellent sense of smell – they can smell a dead animal from four miles away. Their eyesight is not great, though they can see underwater. They can hear, though reptile ears are sensitive to a narrower range of possible sounds compared to mammals.

Octopuses

Humans and octopuses share a common ancestor, deep in the history of our evolution, at least 700 million years ago. Octopuses are part of the molluscs order, the same order as clams and snails. However, their class, cephalopods, which includes squid and cuttlefish, have freed themselves of shells and grown very interesting and very different brains, and developed very complex behaviour, including being incredibly efficient predators. It has been suggested that if we ever want to interact with alien intelligence, we could start by looking at the brains and behaviour of octopuses. The evolution of cephalopod brains has a completely different profile from that of the brains found in vertebrates, and it's extraordinary to consider the differences between brains this has led to.

Humans and other vertebrates have a brain enclosed in a skull, and the brains communicate with the body through the spinal column. Octopuses have about 500 million neurones (far more than any other invertebrate) and nine brains. They have one central brain, which surrounds their mouth (in common with all molluscs), and they have a brain at the base of each of their eight tentacles. These tentacle brains enable the tentacles to be operated independently if required, and also process information from the tentacles, which can sense touch, taste and possibly also visual information. This complex brain system, combined with their highly plastic bodies, means that octopuses are capable of tremendous cognitive and physical flexibility, allowing them to solve problems and explore their worlds in a highly complex way. There are also many reports of octopuses interacting socially with humans, including displaying tactile and playful behaviour, so maybe we are as interesting to them as they are to us.

Insects/crustaceans

It's tempting to dismiss many invertebrate brains as less interesting – though Darwin thought that the brain of an ant was 'one of the most marvellous atoms of matter in the world', able to produce complex behaviours in such a tiny animal. At the other end of the invertebrate size scale, we find the mantis shrimp. The mantis shrimp is a large crustacean that lives in crevices and is a slightly terrifying-sounding predator, able to either smash or spear its prey. They are tremendously strong and have been known to bash their way through the glass walls of an aquarium. They have very complex visual systems. Many animals, including insects, cephalopods, reptiles and birds, can see polarised light – the directionality of the waves in light rays – but the mantis shrimp is the only animal we know of that can see circular polarised light – where the light waves spread out like a spiral. This is reflected in the structure of their very complex eyes, and also in their bodies – the mantis shrimp is the only animal whose surface reflects light that has circular polarising propagation. This strongly suggests that this ability is there for the detection of other mantis shrimps – possibly to avoid a fight! Looking at their nervous system, mantis shrimps are also unlike other crustaceans in having mushroom bodies. Mushroom bodies are clusters of nerve cells found in the brains of insects, which have important roles in behaviours like learning and short-term memory. Controversy exists about what this means – the more complex brains are consistent with the more complex behaviour of the mantis shrimps, but mushroom bodies have been assumed to have evolved after insects split off from crustaceans about 480 million years ago. Their being found in mantis shrimps suggests that mushroom bodies predate this or that there has been some convergent evolution in the mantis shrimps and in a few other crustacea, such as hermit crabs, which also show more complex behaviour, having to travel long distances to hunt. In insects, mushroom bodies are strongly linked to the sense of smell, a sense that is critical to insects, as many of their responses are driven by pheromones (molecules that drive behavioural responses).

Sharks

Sharks are extremely ancient vertebrates, who show a great deal more complexity in their behaviour than was depicted in the film *Jaws*. Shark brains show some of the features we find in our own brains – there is a brain stem-like structure coordinating sensation and action, a cerebellum and a relatively large and rather complex olfactory bulb. Sharks rely a great deal on smell for hunting, and their vison can be relatively poor. Sharks also provide an interesting example relationship between brain size and maternal investment. Found throughout nature, this relationship describes the way that species where the mother spends more time caring for offspring tend to have larger brains. Thus, sharks that lay eggs – with effectively zero maternal involvement or care after that – have smaller brains than sharks that bear live young. Not only are the brains of these sharks larger, but they are also more complex, with a 'forebrain' which is effectively a forerunner of the cortical structures that we find in mammals. Do note: shark mothers that bear live young do not provide care after the baby sharks are born: their maternal involvement is all spent in their time carrying the babies. Great white sharks are 'pregnant' for eleven months!

Birds

Like alligators, birds are reptiles, but we start to find even more complexities in their brains. First, their brains are proportionally larger, relative to their body size, than the brains of other reptiles. There are other differences: compared to other reptiles, birds have a proportionally bigger cerebral cortex and cerebellum. They also have more of their brains dedicated to visual processing than other reptiles, and smaller olfactory areas. This suggests that smell may be less important to birds. Some birds have even bigger cerebellums – in birds like parrots, this seems to be linked to their abilities in manipulating objects with their claws and beaks. Hummingbirds also have very large cerebellums, and this seems to be related to their ability to hover, and to stabilise their eye gaze while they hover, so that they can eat.

Some birds, like parrots and songbirds, are also excellent vocal learners. Like human babies, infant songbirds acquire their vocal repertoires from the adults they hear when they are growing up – though unlike song birds, humans retain the ability to learn new vocal skills for their whole lives. This ability to learn new vocal performances is, in fact, a very rare skill in nature – wherever an alligator grows up, it will make the same sounds. In addition to some birds and humans, vocal learning is also found in some seals, whales, dolphins, bats, elephants and maybe goats and mice. There is a strong genetic component to vocal learning, with the same gene (FOXP-2) implicated in many of these examples. FOXP-2 is expressed in different brain areas in birds and humans, but it seems to underlie vocal learning in both, despite their huge difference in evolutionary terms.

Elephants

Elephants are highly vocal, making an extraordinary range of sounds (and learning how to make new sounds) used primarily to communicate with other elephants. They also have trunks, which are highly extended top lips, including nostrils. These extremely mobile and flexible trunks evolved largely to root for food on the ground, but elephants also use them like human hands, or a mouse's whiskers, continually using them to explore their environment. A bit like us hugging each other or holding hands, elephants also use their trunks to make contact, express affection with, and guide each other. This vocal and tactile contact really matters – elephants live in complex social groups, and they learn both the affiliations and conflicts that they have with other groups, as well as detailed knowledge of their environment (such as where water holes are). All of this knowledge – social and spatial – is essential for the survival of the groups.

If we look at their brains, African elephants have three times as many neurones as us: however, a stunningly high proportion of these (97.5 per cent) are found in the cerebellum. This may well relate to both the coordination needed for such a large body (elephants are more than three times larger than a human) and because they have to control their trunks. There are also some distinct differences in the elephant brain stem, which may relate to these motor skills, and also to elephants' more complex use of sound. Elephants can hear a wider range of sounds than humans, including very low-pitched sounds that we cannot even hear. Elephants can communicate with these low-pitched sounds which can travel great distances through the air: there is also evidence that elephants can hear these sounds through the ground using their feet. The more complex brain stem nuclei in elephants may reflect their integration of sound detected by their ears and by their feet. Finally, elephants have greatly enlarged temporal lobes, consistent with their detailed social knowledge, as the temporal lobe is important for the perception of social cues (such as vocalisations).

Bats

At the other end of the mammal size spectrum, we find bats, small flying mammals. Many bats hunt in the dark, while flying, and their ability to detect prey in the air (such as insects) and also sense structures in their environment (such as trees and walls) is performed not with vision but with sound. Bats use echolocation – they continually make high-pitched sounds and then listen for the echoes, using those echoes to tell them what is 'out there'. Research on bat brains shows that bats who use echolocation have larger inferior colliculi – a small nucleus in the pathway connecting the ears to the brain, the last point in this pathway before the thalamus. One role of the inferior colliculus is collating and comparing information across our two ears, which is how our brains work out where sounds are coming from. So, when comparing the loudness of sounds across the two ears, the inferior colliculus registers tiny differences in time when sounds hit the two ears. For a bat, which needs to do this very quickly while in flight, a larger inferior colliculus may be essential. Notably, bats who hunt not by echolocation but by sound and vision, like fruit bats, do not show this increase in size.

As a side note, humans can use echolocation as well. If you snap your fingers when you are in two very different kinds of spaces, such as inside a car or in a large room, you will notice that the sounds are very different – and that is due to the echoes. Imagine clicking your fingers, or making a clicking sound with your mouth, all the time and using that to navigate! Some people who are blind do actually do this, making clicking sounds to explore their physical surroundings. Brain studies show that this is supported by brain areas that would support visual processing in people with sight – another example of brain plasticity.

Platypuses

The platypus is a monotreme – a semiaquatic egg-laying mammal, with webbed forefeet, a dense fur and (in males) venomous spurs in the hind legs. Platypuses diverged from other mammals 166 million years ago. Their brains (like those of marsupials) have cerebral hemispheres, like other mammals, but they are not joined by a corpus callosum. Platypuses hunt underwater, which raises the question of what information they use to guide this, as they close their eyes and nostrils when they dive. They have a bill, much like that of a duck. The bill contains mechanoreceptors – just like the ones in your skin that you use to detect touch – which are used to detect mechanical vibrations in the water, caused by the movement of prey. The platypus bill also contains two kinds of electroreceptors, which are able to detect electrical activity. The electrical activity they sense is that produced by the muscle activity in their prey. As it swims, the platypus moves its head from side to side, which helps it use this sensory information to zero in on its prey. A great deal of the platypus's brain is devoted to processing this information from the bill, which is also used as a tool to root out food from the riverbed.

Dogs

If we want to see brain adaptations in action, it is helpful to look at dogs. Dogs were the first animals to be domesticated by man, probably starting at least 15,000 years ago. Dogs evolved from wolves, who, like humans, are intensely social mammals, and are cooperative hunters. A long-term study in Russia that started in 1958 has been trying to explore this evolution by selectively breeding silver foxes. The study bred foxes which were happy to be handled by humans separately from foxes that were aggressive towards humans. This led quite quickly to two different kinds of foxes, and the tame 'domesticated' foxes showed some very familiar dog-like responses, such as tail wagging and licking. A brain study recently showed that the tame foxes had increased grey matter, compared to the non-selectively bred foxes, and this was most marked in the prefrontal cortex, the amygdala, the hippocampus and the cerebellum. Strikingly however, these effects were also seen in the foxes bred to be aggressive to humans. This suggests that any selective breeding for behaviour towards humans may be associated with increased brain mass, especially in fields associated with action control, memory and emotion.

There are also many studies now scanning the brains of domestic dogs. One study comparing breeds who have been bred for more specific purposes shows that there are significant differences: the brains of dogs bred for hunting are different from those of dogs bred for guarding, or for herding. And if we look at brain activity, we also see some significant similarities with humans. Dogs show sensitivities to the emotional tone of human voices in similar areas of the brain to us, and they show a sensitivity to familiar humans in similar visual areas of the brain. Perhaps the domestication of dogs really has been built on similarities between our brains and how they work.

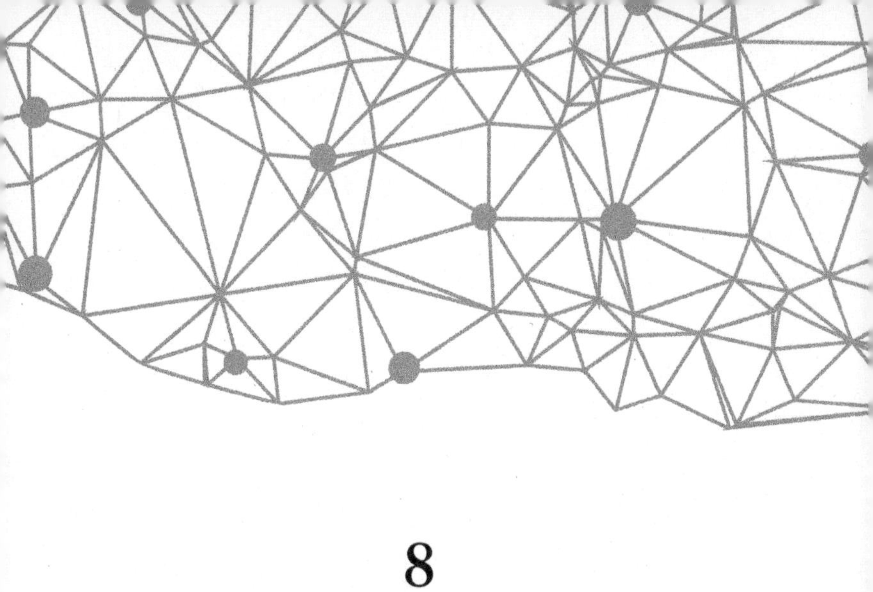

8

How Our Brains Change With Age

We saw in the first chapter that humans have around 86 billion neurones, that we are born with the vast majority of these, and that, along with the rest of our bodies, these neurones are grown during gestation. This results in human babies being born with enormous brains compared to other mammals. In fact, human babies are born about as early as they possibly can be, partly because they are so metabolically demanding on their mothers, and partly because if they were to grow much larger, they could not be delivered by human females, whose pelvises are limited in width by our upright gait. Indeed, as it stands, these large brains already make childbirth exceptionally difficult and dangerous for human females compared to most of our other mammal relatives.

So, human babies have big brains, and these big brains start to grow very fast, quadrupling in size between birth and the age of six, by which time the brain is roughly 90 per cent the size of an adult brain. If we already have almost all the neurones that we will ever have when we are born, what is driving this pattern of growth? It turns out that this increase in size is underpinned by some complex changes that are occurring inside the brain.

We met the structure of neurones in the first

chapter: neurones are formed of a cell body, with a long projection (an axon) and branch-like shorter projections (dendrites) from the cell body or from the far end of the axon. The axons can be thought of as ways the neurone can connect to more distant neurones, while the dendrites connect to nearby neurones. These connections are called synapses. Changes in the brain – associated with learning and development – occur largely through the connections between neurones, which can be through the strengthening of existing connections, or through the development of new dendritic connections. The axons are coated in a slim, fatty sheath called myelin: this enables the electrical discharges that allow transfer of information in the brain to be propagated rapidly along the length of the axon. Although any one dendrite or one myelin sheath will be completely tiny, as these processes develop over 86 billion neurones, the brains continue to change in shape and size.

At birth and in early infancy, many dendritic connections exist, and many more are created between neurones: this is known as synaptic exuberance. In the early years of life, these are rapidly pruned, at first quickly, then more slowly. The implication of this is that much childhood brain development consists not just of forming the brain connections that we need, but of our brains getting rid of connections that are not needed. This seems to match some of what we know about how babies learn. For example, babies are sensitive to many more acoustic features of human voices than adults are. As babies learn about the language(s) that they hear, they become more sensitive to the acoustic features that are relevant for the language they are learning, and less sensitive to the sounds that are not (language varies a lot in this). Maybe the decreased sensitivity to irrelevant sounds is correlated with a loss of the synaptic junctions that once supported it.

During adolescence, a more adult profile of synaptic connections starts to appear; however, this does not appear in a consistent way throughout the whole brain. The brain areas associated with sensory processing show this adult profile first, while the prefrontal cortex shows it last: indeed, even by the age of eighteen, the prefrontal cortex of the adult profile is still not fully established. The relationship between synaptic exuberance and pruning, and their implications for the developing brain and experience are still being explored, but in terms of brain connectivity, the adult pattern is not yet established at eighteen.

The myelin sheaths on our neurones are formed as the brain develops: when we are born, not all our neurones are fully myelinated. Myelination is a process that increases the speed and efficiency of neural function. Myelination in the human brain begins in visual brain areas a couple of months before birth, and continues in other sensory brain areas over the first year of life. This process continues in other cortical and subcortical systems into the middle of the third decade. In other words, myelination continues into our mid-twenties.

The pattern of myelination has been expressly linked to the development of cognitive skills in children and adolescents, as myelination greatly improves the speed of the electrical signalling of neurones, and hence how fast and efficiently they work. Myelination proceeds in a roughly back-to-front direction in the brain, from visual cortex to prefrontal cortex. This means that the frontal and prefrontal fields are the ones that are still being myelinated into our mid-twenties, and so until we reach this age, the connections to the frontal lobes are not myelinated like a mature adult brain. This is likely to be most obviously related to experiences and behaviours associated with the prefrontal cortex, functions such as decision-making and judgement tasks.

One might imagine that the development of the brain in childhood and adolescence would progress in a relatively linear way. However, both synaptic connections and myelination progress in an uneven ways throughout the brain. If we look in detail at the structure of the brain from birth to the age of eighteen, we see a non-linear pattern of change in the proportion of white and grey matter, which may partly involve changes in myelination (see above), and also appears to involve the loss of cells through cell death. A recent study looking at this pattern into adolescence found that, while brains increase in volume throughout adolescence, the volume of grey matter is at its highest in childhood. This then decreases (proportionally) throughout our teenage years, while white matter *increases* (proportionally) through our teenage years. This indicates that considerable changes are still happening in the structure of the adolescent brain. In terms of specific brain areas, while the cortex continues to thin throughout adolescence, the decreases are most marked in the parietal lobes, and least marked (or growth is seen) in temporal and prefrontal fields.

Again, this pattern of brain development seems to suggest that our brains don't reach an adult profile of structure until our twenties. The pattern of maturation of the brain in adolescence suggests a particular issue with frontal-lobe functions, with the frontal and temporal lobes showing a different pattern of change (in terms of movement towards adult profiles) compared to parietal and visual brain regions. In addition, as we've seen, the frontal lobes are the last to be fully myelinated. The frontal lobes are associated with complex cognitive control processes: so-called 'meta-cognitive processes' that enable us to plan our behaviour, control our responses, adapt our behaviour to different contexts and requirements, and anticipate the implications and consequences of our behaviour. The absence of mature frontal-lobe connectivity and functions has been linked to increased impulsivity and risk-taking in adolescence, and to greater susceptibility to peer opinions and behaviour.

It's tempting to imagine that we hit our mid-twenties and then our brains settle into an adult profile, anatomically, from which we trundle towards old age, when everything starts collapsing. However, this is not quite true: as we will see in the next chapter, our brains are highly changed by experience. Anything you learn, any new memory, every new skill: all of this is reflected in changes to your brain. One of the tremendous abilities of humans is their terrific inventiveness and their flexibility in adapating to different environments – and all of this relies on our brains' abilities to change. The adult brain state is one of continual reinvention and change.

However, these adults are all still getting older. What do we know about the anatomy of the aging brain? Structural imaging of healthy older adults between the ages of sixty-three and seventy-five shows that there are some marked changes that correlate with our ages. As we get older, there are decreases in the volume of both grey matter and white matter. Cerebrospinal fluid volume significantly increases with age; the brain floats in cerebrospinal fluid, and as the brain gets smaller, the volume of cerebrospinal fluid increases. A specific analysis of the different brain regions showed that the highest rates of grey matter atrophy in this age range were localised in primary cortices, including the primary auditory cortex, the primary visual cortex, the primary somatosensory cortex, and the primary motor cortex. Rate of loss was also high in the parietal cortex, the orbito-frontal cortex and the hippocampi. This is interesting as it – to some degree – follows the progression of brain development in infancy and childhood. The prefrontal cortex is the last to show an adult profile, and also one of the last areas to show the effects of old age. In many other areas, the rate of change with age is minimal, while in others – such as the basal ganglia – there is an increase with age.

Finally, if words like 'highest rate of loss' are scaring you, then it is important to note that the highest rate of change that correlated with age was 0.83 per cent per year – that is, less than 1 per cent change per year – and in much of the brain, the effects are even smaller. So we are not talking catastrophic effects on your brain.

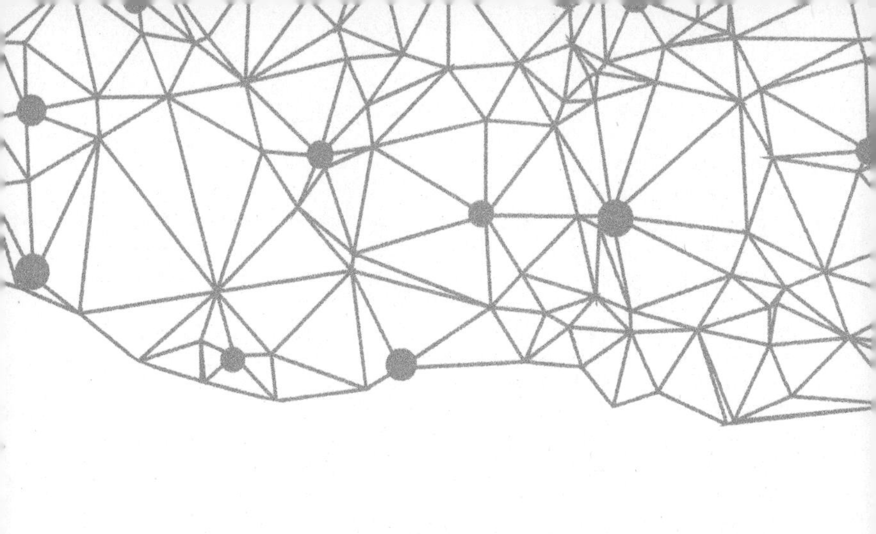

9

How and Why Are Brains Different From Each Other?

Your brain is different from my brain, and from everyone else's, in terms of the precise anatomy, the information it has encoded, and in aspects of its function. This is because human brains are affected by our experiences and also by some of the characteristics we are born with – our genetic heritage.

Biological sex

First, the human brain is affected by our biological sex. Unlike our bodies, having a look at anyone's brain does not give any information about the biological sex of the owner of that brain. There are, however, striking effects of sex on the size of the brain: male brains are larger than female brains, even when body size is accounted for. In contrast, female brains have a relatively thicker layer of grey matter than male brains, and this is often expressed as a higher proportion of grey matter to white matter. You will recall that grey matter forms the surface of the human brain, and it contains the cell bodies of the neurones (neurones), while white matter is made of the long projections of neurones, connecting different brain areas. The greater proportion of grey matter in females may imply that while female brains are smaller overall, they are fitting comparably similar amounts of tissue to support computational processes into this smaller space. Of course, this does not tell us why male brains are bigger – if it is not to fit in more computational power, then we need another explanation. It's also worth bearing in mind that human brains are extremely expensive metabolically, so those larger male brains come at a metabolic cost. Also, as noted earlier, human brains use around 20 per cent of the circulating oxygen in our bloodstream, as they are large and the energy required to maintain the potential for the neural systems to work is considerable. As we've seen, the large heads that these large brains need make a big contribution to the considerable physical danger that childbirth constitutes for human females – globally, around 800 females die in pregnancy and childbirth every day. So our big brains – male and female – are expensive, and they need to pay their way.

IQ

Another significant source of variation across humans is overall cognitive ability, often called IQ (intelligence quotient). A recent study of thousands of people in the UK showed that there are significant correlations between IQ and brain structure. These were found in very distributed brain regions, including the insula, frontal lobes, the front end and top of the temporal lobes, hippocampal areas, visual areas, and the thalamus. It's tempting to interpret this as showing the brain regions that drive intelligence, but of course we don't know how these brain regions have been affected by development: does the brain drive IQ, or does IQ affect how the brain develops? We know from genetic studies that IQ is partly heritable, but is also affected by environments and opportunities – and these probably also affect the resulting pattern of brain anatomy. Strikingly, the effects in this study were much clearer in older participants, suggesting that life experience is affecting the results.

Personality

Humans also vary a great deal in their personalities. You may well have done personality tests for work or for fun: the classic and very reliable 'big five' personality factors are extroversion/introversion, neuroticism/stability, openness, conscientiousness and agreeableness. It has been difficult to identify brain areas that correlate in structure with different personality traits according to the big five. Further research looking at functional connectivity in the brain (how different brain areas correlate with each other in their activity) shows some differences that correlate with personality measures. It is probably fair to say that personality can make people very different from one another, but we may not yet have established the best ways of capturing how this relates to brain structure and function.

Handedness

Humans are unusual in that not only do we have very mobile and flexible control of our hands, we also show a distinct preference for one of our hands when we do things. This is normally called handedness, and for the vast majority of people, the preferred hand is the right hand. Around 7 per cent of people show the opposite preference, using the left hand. This profile is much harder to find in other animals, who may show an individual preference for one hand (or claw), but not this marked right dominance over the whole population. Many have wondered whether there are distinct brain differences between left- and right-handed adults. It turns out there is very little evidence for cortical differences between right- and left-handers, but there are possible differences in the basal ganglia. These regions are important in motor control, and suggest that brain differences between left- and right-handers have more to do with the different motor-control challenges than cognitive differences in brain function.

So we can see that our brains do vary, and this can be associated with aspects of our genetic inheritance. But we have also seen that human brains don't work 'out of the box' – we have a very long period of brain development in childhood and adolescence. So how are our brains affected by the cultures we grow up in and the experiences we have?

Spoken language

When we listen to spoken language, we see activation in the left hemisphere associated with the perception of different linguistic elements in speech: the speech sounds, the syntactic information, the semantic information. This engages left temporal-lobe regions, and extends out into frontal areas associated with the control of speech production, and a wider left-hemispheric network of language related areas. As we do this, the right brain is also recruited, often associated with other kinds of information about the speaker, such as their identity and the musical intonation of their voices. If, however, we consider a language like Mandarin Chinese, the music of someone's voice directly influences what they are saying, as Mandarin is a language that uses what is called lexical tone. The Mandarin word '*ma*' can be said in five different ways, with five different intonations, and five completely different meanings. When Mandarin speakers listen to spoken Mandarin, they show equal amounts of left and right temporal-lobe activation, arguably because they need to recruit the musical processing of intonation into the language systems in the left hemisphere.

Strikingly, when people who are deaf and use sign language to communicate are in brain scanners, they show visual-cortex activation when they see someone signing, but their brain activation then follows the same patterns of left temporal-lobe activation as is seen when hearing people listen to speech. So both signed and spoken languages look very similar in the brain.

Reading

Modern humans emerged between around 100,000 and 200,000 BC. However, humans have only been reading and writing for 6,000 years, and our brains have, arguably, not yet adapted to this skill. Unlike spoken language, which babies learn without overt instruction, we typically must be taught to read, but by the time we are adults, most people are such skilled readers that they can read highly automatically with no sense of effort. Research suggests that learning to read changes the responses in existing brain networks, often overlapping with language and face-processing areas. Consistent with this, children learning to read English and Chinese, two very different writing systems, show similar changes in the visual cortex as their brains become more proficient at reading.

Reading systems also differ: for example, Italian is a 'transparent' writing system – letters always map onto the same sounds. English is highly irregular, as the same letters can be said in many ways (compare the '–ough' sequences and the sounds they map onto in 'through', 'tough', 'though' and 'thought'). This leads to brain differences when people read in the two different languages – people reading Italian show activation in the temporal lobe, while people reading English show this, plus more activity in frontal areas associated with speech production, suggesting that the English readers need to do more work to map letters onto speech sounds.

Finally, just as signed languages look very similar to spoken languages in the brain, when braille readers read text with their fingertips, we see activation in the somatosensory cortex, and the activation then runs down into the same visual areas that process visual-text, meaning visual- and braille-reading systems both look very similar in the brain.

Bilingual brains

Many people speak more than one language – how does this affect our brains? This seems like a simple question, but it has proved to be difficult to answer definitively, as people learn second languages (or more) at different ages and for different reasons – and, of course, the languages themselves vary. By many definitions of bilingualism, I am bilingual, as I speak some French; however, my French expertise is nothing like that of a child who has grown up in a household where both English and French are spoken. That being said, a recent analysis of many studies about bilingualism or multilingualism found it has many significant brain effects in the frontal lobes, the anterior cingulate cortex, left inferior parietal lobule and subcortical areas. What is striking here is that while some of these areas are associated with spoken-language processing (e.g. the inferior parietal lobes), the effects also extend beyond this, into the frontal lobes and the anterior cingulate. This is consistent with suggestions that the demands of having to switch between different languages mean multilingual people are better at attentional tasks. Other than this, the study found that multiple languages are all represented in similar brain regions – they are not stored away separately.

Learning more than one language has been suggested to provide benefits against dementia (in addition to being a useful skill in and of itself). Sadly, the picture here is still not clear – being multilingual probably does not reduce your risk of developing dementia. It is also the case that at a global level, it is normal to be multilingual – around the world, most people speak more than one language. It is highly monolingual environments like the UK that are the exception, so we might not expect multilingualism to be an advantage – it may simply be the norm for humans.

Music

How does learning to play a musical instrument affect the brain? Musical expertise includes learning to read music, learning to make new complex actions, and learning to listen to the sounds we and other musicians are making. All these elements appear to result in differences in the musician's brain. If we directly compare musicians and non-musicians, we find that musicians show more grey matter in the primary auditory cortex, potentially associated with more complex listening skills, and more grey matter in frontal lobe regions associated with executive control processes. More grey matter is also found in the hippocampi, associated with memory processes, the lingual gyrus, linked to reading musical scores, and the primary somatosensory cortex, associated with sensory motor control. Similarly, when we look at brain function, musicians show very different brain activity when they listen to compared to non-musicians. However, as with the IQ study mentioned above, we cannot be sure if this reflects the effects of musical training, or a brain predisposition that leads people to engage with music. There are also big social influences on whether a child gets the opportunity to learn a musical instrument at all.

Neurodiversity

A final way that humans can vary refers to the idea that there are some fundamental differences in how our brains develop – you may have heard this called 'neurodiversity'. In a more general sense, neurodiversity can be used to describe brains that differ from the 'norm'.

To tackle neurodiversity properly would take a

whole new book; there are many different developmental trajectories that can affect people in different ways. For example, in dyslexia, children can have severe issues learning to read and spell. In autism, children can struggle with understanding social situations and with communication. The current understanding from a brain perspective is that these neurodiversities are frequently associated with differences in the connectivity of the brain. For example, people with dyslexia show a distinctly different profile of connectivity between speech perception and speech production networks – and learning to read relies heavily on these networks. In response, their brains might find a different way of solving the problem of how to read, often by a different route. This can often mean that reading and spelling remain effortful and error-prone. The brains of people with autism also show considerable anatomical differences, including smaller amygdala, decreased grey matter in the bottom of the temporal lobe, and increased cortical thickness in the frontal cortex. There seems to be a strong effect of age interacting with this profile, with the largest differences occurring around adolescence.

Of course, all of us have brains that reflect our age, our biological sex, our education, the languages we speak, the music we play, and the ways we fall on different neurodiversity spectra. And these can intersect: for example, my dad was from the generation where teachers forced left-handed children to write with their right hands; he also left school at fourteen and started working. As a result, he struggled with handwriting for his whole life. His adult brain probably would have been very different if he'd been allowed to write with his left hand, and if he'd had a chance to go to college. But he was a voracious reader and a skilled musician, and his brain will have reflected these skills as well. And this is true for all of us. Probably the best approach for our brains is to give them as many opportunities for development and growth as possible.

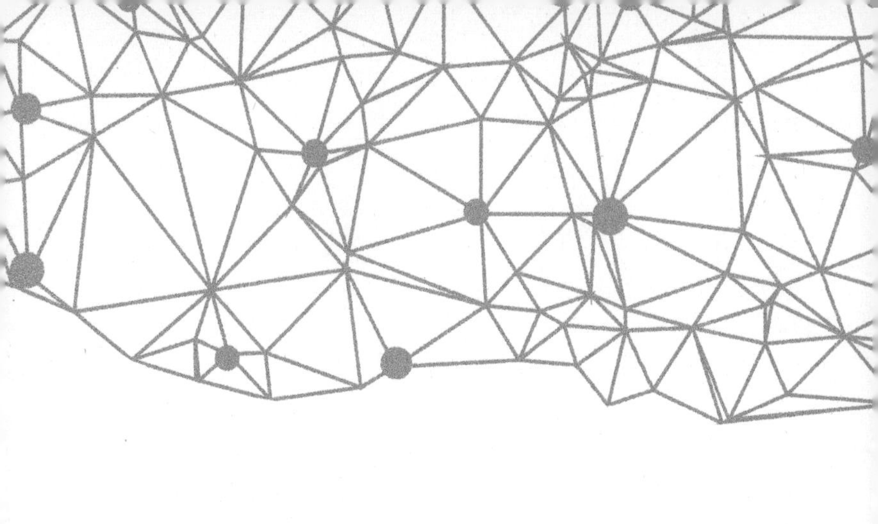

10

What Can Damage the Brain? What Is Good for the Brain?

The brain is a large and highly complex organ, and unfortunately can be damaged in many ways. Typically, it is the brain areas affected by the damage that will determine the symptoms seen, although the severity of the damage will also influence recovery.

The brain consumes a great deal of the oxygen that circulates in our red blood cells. As noted in chapter 9, this is because our brains use a lot of energy simply keeping neurones primed and ready to fire. In turn, this means that our brains are extremely dependent on the health of the cardiovascular system – our hearts and our blood vessels. Indeed, disorders involving the blood supply to the brain are the most common cause of brain damage in our culture. When there is sudden disruption in the blood supply to the brain, this is called a cardiovascular accident, or a stroke.

Strokes have three main characteristics. The first is that the symptoms of brain damage start abruptly; these can indicate quite specific symptoms– for example, slurred speech or arm weakness. The second is that the problems associated with the stroke are generally most severe at the onset of the stroke (or very soon after). Finally, if the person suffering a stroke survives, there is frequently, over a longer timescale, some improvement in the problems they experience.

This means that there is some possibility for the recovery of some functions, as the brain reorganises around the damaged area.

There are two different kinds of stroke. An ischaemic stroke happens when there is not enough blood supplying certain parts of the brain for cells to survive. This can be caused by a blood vessel completely closing or by something blocking the blood vessels, such as a blood clot. The problems associated with the total narrowing of the blood vessels are the most common causes of strokes, and they can develop quite suddenly, often during sleep or shortly after getting up (because our blood pressure drops during sleep, which can cause the blood vessels to narrow further).

Both forms of ischaemic stroke deprive the brain of oxygen, leading to the death of cells in the middle of the affected brain areas. Around the focus of the stroke, there can be areas where the cells are disrupted by the oxygen deprivation, but do not die. Changes in the activity of these cells is one of the ways that the brain can start to show recovery after a stroke: the other ways that the brain recovers involves the rest of the brain remodelling its connections around the damage.

The severity of these attacks can vary enormously. If a major blood vessel is blocked, the effects in the brain can be serious and widespread. Alternatively, the oxygen deprivation can be brief, and may not even last long enough to cause brain damage. For example, transient ischaemic attacks (TIAs) are brief episodes of focal neurological problems (e.g. lack of sensation in the fingers), followed by complete recovery. TIAs are not serious in themselves, but they need to be checked out, as they can suggest someone is at greater risk of a stroke.

The second kind of stroke is a haemorrhagic stroke, which happens when one of the blood vessels in the brain is breached. This can start abruptly, or may gradually worsen over several hours. These strokes often occur when people are active, and are not necessarily preceded by any warning signs, although the affected person may have a severe headache, start vomiting, or start to lose consciousness. These strokes can be caused by high blood pressure and points of weakness in the walls of blood vessels, or precipitated by head injuries. They are often more serious than ischaemic strokes, but again, if the affected person survives, some recovery is possible.

Brain tumours – a growth of abnormal cells in the brain – are a less common disorder of the nervous system. Tumours can be either benign or malignant, and both can affect the brain. The brain sits within the skull, a fixed space, so growing tumours increase pressure on the brain, which can compress and distort the brain tissue. Tumours can also affect the flow of blood through blood vessels to various brain areas. Tumours may cause specific damage to brain areas, resulting in symptoms such as weakness in arms or legs, or epileptic fits.

Damage to the head is another common cause of brain injury. Open head injuries ('open' meaning that the brain is penetrated or exposed by the injury) are obviously very severe, but closed head injuries (a traumatic injuries that do not expose or penetrate the brain) can also be highly serious. Closed head injuries can cause bleeding in the brain (i.e. a haemorrhagic stroke), and they can also lead to lasting and pervasive brain damage. This is because the brain floats in the cerebrospinal fluid inside the skull. Sudden rotations or impacts of the head can cause the brain to move – for example, being twisted around the brain stem, which can directly damage the brain stem. The brain can also be struck against the inside of the skull, and brain tissue can be damaged by shearing and twisting. All of this means that closed head injuries can lead to diffuse damage to the brain. This in turn can lead to persistent changes in behaviour, including difficulties understanding the implications of what someone is saying, or difficulties controlling actions and emotions.

In our Western culture, head injuries are common in two very different populations: young adult males, and older adults. The most common cause of head injury in young males is road traffic accidents. The most common cause of head injuries in older adults are falls: humans are very unusual in that we walk upright, and our bipedal gait is a form of controlled falling. When we are younger, the inherent instability of walking on two feet isn't really a problem, as our brains and bodies are on the case, but as we get older, brain and muscular changes can make us less steady on our feet and more prone to falls.

Degenerative disorders are another cause of brain damage. 'Degenerative' means brain damage that is characterised by a progressive worsening of brain function, and atrophy, or shrinking of brain tissue, caused by the loss of neurones. You will more commonly hear this called dementia, although the type of degeneration that someone is coping with can be very different depending on the precise kind of dementia that is affecting them. Alzheimer's disease is the most common form of dementia – around two-thirds of people with dementia have Alzheimer's disease (although many will not have a formal diagnosis). There are some distinct brain changes associated with Alzheimer's disease, including a build-up of proteins in neurones that may impair their function. The first symptoms can be quite mild, and often reflect memory problems, which are usually associated with damage to the hippocampus and the surrounding brain areas. The actual progression of Alzheimer's disease can be very variable, depending on which brain areas become affected: people with Alzheimer's disease can have many different problems with aspects of daily activities, and will require daily care. The main risk factor for Alzheimer's disease is age – it becomes much more common over the age of eighty. A second risk factor is sex – women are more at risk than men.

Pick's disease, also known as frontotemporal dementia, resembles Alzheimer's, but is far less common, and tends to start with focal atrophy (shrinkage) of the anterior temporal lobes or the frontal lobes. Depending on where this happens, the first problems can be seen in changes in ability to understand language, or changes in behaviour or personality. Frontotemporal dementia affects people at a younger age than Alzheimer's disease, and affects more men than women.

Multi-infarct dementia is also relatively common, and results from someone suffering multiple small strokes. The progression of this kind of dementia can be 'step-like' rather than gradual, as this reflects the occurrence of small strokes, and there may be some improvements in function between the 'steps'.

Other degenerative diseases are linked to changes in the basal ganglia, such as Parkinson's disease and Huntington's chorea. Huntington's chorea is an inherited disorder that can lead to emotional changes, depression and problems controlling movements. Parkinson's disease primarily leads to issues with movements, such as difficulty in initiating actions, very slow movements, tremors, rigidity of muscles and difficulties in producing facial expressions.

It's alarming to think of the many ways in which things can go wrong with our brains, and it's also worth thinking about the many people who are coping with these patterns of brain damage, as well as all the other people affected by this: their families, their carers. People are dealing with a lot of very challenging circumstances. However, it is also important to know that the situation is not entirely negative: when Italy introduced mandatory crash helmets for people riding motorcycles and mopeds in 2000, accident rates were unaffected, but head injuries dropped by 66 per cent. Medical treatments for Parkinson's disease can be highly effective, and there is very promising work underway looking at genetic methods for treating Huntington's chorea. People can suffer greatly with Alzheimer's disease, but can still enjoy familiar music, which seems to be able to help them access emotions and memories via routes unaffected by the disease. And there are some other positive steps that we can take to improve our own brain health, which may have some preventative or neuroprotective effects.

Being active

Scientific research over the past decade has shown that moderate, regular activity has significant neuroprotective effects on our brains. Anything that can improve our cardiovascular fitness will have benefits for the brain, as the brain is so dependent on our hearts and our blood vessels bringing oxygen to it. I know at this point you may feel tempted to throw this book across the room, because you are sick to death of people telling you that exercise is good for you, but it's important to remember that exercise is not only for incredibly fit people, and that when it comes to activity, every little helps. So, although I am unlikely to ever run a marathon, I can still go out for what may well be the world's slowest jog, and that still counts. In fact, every step counts. I think of daily exercise as a brain hack that improves my mood in the short term (all those endorphins) and my brain health in the long term. Moderate activity and exercise are linked to lower rates of strokes, and are also associated with lower rates of dementia. And it's never too late to start.

I have mentioned at several points that we are born with almost all the neurones we will ever have. It used to be thought that we could never grow new neurones in our central nervous system. However, there is now good evidence that (certainly in rats) new neurones can be grown in the hippocampus – and this is initiated by exercise. So, your exercise could even grow some new neurones! I do also have to say, pick an activity that works for you, and contact your GP before you start if you have not been active for a while.

Other factors that can affect your cardiovascular system in a negative way include smoking, drinking more alcohol than is recommended, and not maintaining a healthy weight. (You have my full permission to throw the book across the room again now. Going and picking it up will count as activity!)

Balanced diet

You will recall that all the neurotransmitters your brain uses are synthesised in the brain itself. You may read about 'dopamine diets' or other ways of managing your food intake in a way that means you may be able to alter your brain chemistry. However, in truth, a healthy, balanced diet will provide you with all the nutrients you need to make those lovely brain chemicals. The phrase: 'Eat food. Not too much. Mostly plants' is quite a good starting point.

Brain training

You will probably have heard about brain training – the idea that doing puzzles or computerised tests, or something that makes your brain work a bit harder, is good for it. It's certainly not *bad* for your brain – and you will definitely get better on those tests or those puzzles, which is rewarding in itself. So, if you enjoy games and puzzles like this, do not stop doing them! However, the evidence that playing, for example, a word search game will make you better on cognitive tasks that are not word search games is harder to come by.

In chapter 9, we saw that things like musical training or learning to read really do affect brain structure and function. The difference with brain training is likely to be the sheer amount of time spent doing it. One reason why it's harder to learn a spoken language when you are an adult is that when you do it as a child, you spend a huge amount of time on it – time adults often do not have. It's striking that some studies find that people who play a great deal of video games often score more highly on tests of visual attention, which reflects the demands of the games, and also the sheer amount of time they have put into this, time that might otherwise be spent doing other activities, such as being physically active.

Social contact and hearing aids

Loneliness is very bad for our physical and mental health – humans are social primates, and the size and connectivity of our social networks predicts our risk for physical diseases, mental health issues and even our longevity. This is partly because friends can be practical sources of help and support, but it's also because we get positive mental benefits from social contact – having conversations and laughing with people increases endorphins, and reduces adrenaline and cortisol levels, so we are happier and more relaxed. We are also exercising our brains when we have social interactions with other people.

And this may lead us to the root of why the single biggest preventable risk factor for dementia is uncorrected hearing loss in adulthood. Losing your hearing in adulthood and not using hearing aids to support your hearing increases your dementia risk by nearly 10 per cent. Hearing loss is relatively common as we age. If not treated with hearing aids, hearing loss can lead to people finding social interactions difficult, as it is hard to hear other people speak. This, in turn, can lead to people withdrawing from social interactions. And it is this withdrawal from conversations and contact that seems to put the brain in a riskier situation. If we have age-related hearing loss, hearing aids can take a while to get used to, and people can feel negative towards them for emotional reasons. However, much like exercise, we should think of them as brain hacks that can provide really important neuroprotection. If you are starting to lose your hearing, get tested and if you need hearing aids, get them fitted and keep talking to people; it really is good for your brain.

In this book, I have tried to offer a glimpse of the beauty and complexity of brains. We are continually learning more about the variety of brains out there. As neuroscience continues to develop, we are getting a much clearer picture of how and why brains differ, and how evolution has driven these changes in pace with physical changes in bodies and behaviour. Genetic studies help us understand how ancient some of the genetic codes are that underlie the elements of our nervous system, and these are essential to the structure and function of brains. We are getting better and better at studying the wonders of the human brain – with many possible benefits for all of us. We are learning more about how we can protect our brains from damage, and how we can treat our brains when different factors affect them negatively. And there will be so much more to learn about all these dazzling and complicated brains in the future. Where are you going to take your brilliant brain next?

Acknowledgements

I would like to thank all the people who have been kind enough to share their brains with me, especially DR and SE, who were so kind, patient and generous. I would also like to thank all the scientists I have worked with, including Andy Calder, Andy Young, Paul Burgess, Alex Leff, David Sharp, Jane Warren, Jenny Crinion, Carolyn McGettigan, Zarinah Agnew, Narly Golestani, Charlotte Jacquemot, Disa Sauter, Patti Adank, Frank Eisner, Jonas Obleser, Sam Evans, Sinead Chen, Sophie Meekings, Cesar Lima, Nadine Lavan, Kyle Jasmin, Ceci Cai, Alexis Mcintyre, Addison Billings and Caz Niven. I would like to thank John Mellerio from the Polytechnic of Central London for trying his level best to teach me about the brain. I would like to thank Alex Griffin for pointing out how strange it is that your toes are only two neurones away from your brain. I would like to thank Will Eaves for his writing advice and his encouragement, and for being a terrific podcasting colleague (together we are the Neuromantics! Available wherever you listen to your podcasts). I would like to thank Duncan Wisbey, David Arnold and Harry Yeff for having exceptionally interesting brains. I would like to thank Will Francis and my editor George Brooker, for getting this to happen at all. Finally I would like to thank Geraint Rees for letting me steal his alligator story, and I am glad it didn't eat him.